STATE-OF-THE-ART OF QUANTUM DOT SYSTEM FABRICATIONS

Edited by**Ameenah Al-Ahmadi**

State-of-the-Art of Quantum Dot System Fabrications
Edited by Ameenah Al-Ahmadi

Published by InTech

Janeza Trdine 9, 51000 Rijeka, Croatia

Publishing Process Manager Molly Kaliman

Technical Editor Teodora Smiljanic

Cover Designer InTech Design Team

Additional hard copies can be obtained from orders@intechopen.com

State-of-the-Art of Quantum Dot System Fabrications, Edited by Ameenah Al-Ahmadi
p. cm.
ISBN 978-953-51-0649-4

For EU product safety concerns:
IN TECH d.o.o., Prolaz Marije Krucifikse Kozulić 3, 51000 Rijeka, Croatia,
info@intechopen.com or visit our website at intechopen.com.

Meet the editor

Dr Ameenah N. Al-Ahmadi is currently a Professor of physics at Umm Al-Qura University in Makkah, Saudi Arabia. She received the M. Sc. degree in Physics in 2004 and awarded the Ph. D degree in Physics on June, 2006 from Ohio University at Athens, Ohio, USA. Her research interests is in the field of optical properties of nano scale system, including quantum dots, nanowires, nanotubes. She has contributed many papers in different International Journals in her area of research. She has two papers selected by Virtual Journal of Nanoscale Science & Technology in 2004 and 2006. She was the director of enrichment summer program for Giftedness & Creativity_ Sense the word_ 2010. She has been Vice-Dean of Applied science Faculty for the Academic Development and Community Services and graduate student advisor of Faculty of Applied science since 2009 and Vice-Dean of Quality and Academic Development at Umm Al-Qura University.

Contents

Preface

Quantum dots are one of the most promising types of nanoparticles, which are exceptionally useful for variety of new applications because of their unique properties. This is a collaborative book sharing and providing the academic community with a base text that could serve as a reference in research by presenting up-to-date research work on the field of quantum dot systems. We are most grateful to all authors of the chapters for highlighting the important issue of the potential applications of quantum dot system with a high quality work of their research. We are especially thankful for the cooperation and support from InTech team who helped in publishing this book, in particular the publishing process manager of this book, Ms. Molly Kaliman for her hard effort and patience during the process of publishing the book.

"To my son Azuz"

Ameenah N. Al-Ahmadi, PhD
Associate Professor of Physics
Faculty of Apllied Science
Umm Al-Qura University
KSA

1

Synthesis of Glutathione Coated Quantum Dots

Jana Chomoucka[1,3], Jana Drbohlavova[1,3], Petra Businova[1],
Marketa Ryvolova[2,3], Vojtech Adam[2,3], Rene Kizek[2,3] and Jaromir Hubalek[1,3]
[1]*Department of Microelectronics, Faculty of Electrical Engineering and Communication, Brno University of Technology*
[2]*Department of Chemistry and Biochemistry, Faculty of Agronomy, Mendel University in Brno*
[3]*Central European Institute of Technology, Brno University of Technology*
Czech Republic

1. Introduction

QDs play an important role mainly in the imaging and as highly fluorescent probes for biological sensing that have better sensitivity, longer stability, good biocompatibility, and minimum invasiveness. The fluorescent properties of QDs arise from the fact, that their excitation states/band gaps are spatially confined, which results in physical and optical properties intermediate between compounds and single molecules. Depending on chemical composition and the size of the core which determines the quantum confinement, the emission peak can vary from UV to NIR wavelengths (400–1350 nm). In other words, the physical size of the band gap determines the photon's emission wavelength: larger QDs having smaller band gaps emit red light, while smaller QDs emit blue light of higher energy (Byers & Hitchman 2011). The long lifetime in the order of 10–40 ns increases the probability of absorption at shorter wavelengths and produces a broad absorption spectrum (Drummen 2010).

The most popular types of QDs are composed of semiconductors of periodic group II-VI (CdTe, CdSe, CdS, ZnSe, ZnS, PbS, PbSe, PbTe, SnTe), however also other semiconductor elements from III-V group such as In, Ga, and many others can be used for QDs fabrication (e.g. InP) (Wang & Chen 2011). Particularly, much interest in nanocrystals is focused on the core/shell structure rather than on the core structure (Gill et al. 2008). Majority of sensing techniques employing QDs in biological systems are applied in solution (colloidal form). Up to present days, the most frequently used approaches have been reported on the preparation of colloidal QDs: hydrophobic with subsequent solubilisation step, direct aqueous synthesis or two-phase synthesis. Compared with hydrophobic or two-phase approaches, aqueous synthesis is reagent-effective, less toxic and more reproducible. Furthermore, the products often show improved water-stability and biological compatibility. The current issue solved in the area of QDs synthesis is to find highly luminescent semiconducting nanocrystals, which are easy to prepare, biocompatible, stable and soluble in aqueous solutions. Thus, the semiconductor core material must be protected from degradation and oxidation to optimize QDs performance. Shell growth and surface modification enhance the stability and increase the photoluminescence of the core.

The key step in QDs preparation ensuring the achievement of above mentioned required properties is based on QDs functionalization. Most of these approaches are based on bioconjugation with some biomolecule (Cai et al. 2007). Many biocompatible molecules can be used for this purpose; however glutathione (GSH) tripeptide possessing the surface amino and carboxyl functional groups gained special attention, since it is considered to be the most powerful, most versatile, and most important of the body's self-generated antioxidants. GSH coated QDS can be further modified, for example with biotin giving biotinylated-GSH QDs which can be employed in specific labelling strategies (Ryvolova et al. 2011). Namely, these biotin functionalized GSH coated QDs have high specific affinity to avidin (respectively streptavidin and neutravidin) (Chomoucka et al. 2010).

2. Glutathione as promising QDs capping agent

GSH is linear tripeptide synthesized in the body from 3 amino acids: L-glutamate, L-cysteine, and glycine (Y.F. Liu & J.S. Yu 2009) (Fig. 1.). These functional groups provide the possibility of being coupled and further cross-linked to form a polymerized structure (Zheng et al. 2008). Thiol group of cysteine is very critical in detoxification and it is the active part of the molecule which serves as a reducing agent to prevent oxidation of tissues (J.P. Yuan et al. 2009). Besides its thiol group acting as capping agent, each GSH molecule also contains one amine and two carboxylate groups (Chomoucka et al. 2009).

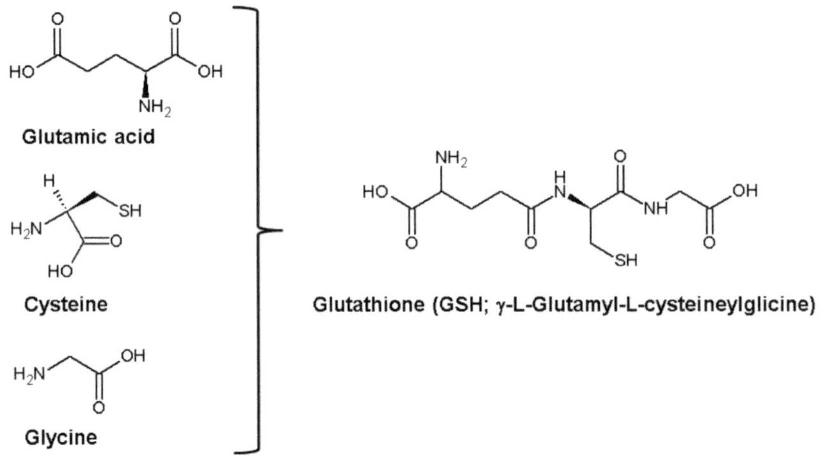

Fig. 1. Structure of glutathione

GSH is presented in almost all living cells, where it maintains the cellular redox potential. The liver, spleen, kidneys, pancreas, lens, cornea, erythrocytes, and leukocytes, have the highest concentrations in the body, ordinarily in the range from 0.1 to 10 mM. It belongs to powerful anti-viral agents and antioxidants for the protection of proteins, which neutralize free radicals and prevent their formation (Helmut 1999). Moreover, it is considered to be one of the strongest anti-cancer agents manufactured by the body. GSH´s important role is also in the liver for detoxification of many toxins including formaldehyde, acetaminophen, benzpyrene and many other compounds and heavy metals such as mercury, lead, arsenic and especially cadmium, which will be discussed later concerning the toxicity level of Cd-based QDs. GSH is involved in nucleic acid synthesis and helps in DNA repairing (Milne et al. 1993). It slows the aging processes; however its concentration decreases with age. GSH must be in its reduced form to work properly. Reduced GSH is the smallest intracellular thiol (-SH) molecule. Its high electron-donating capacity (high negative redox potential) combined with high intracellular concentration (milimolar levels) generate great reducing power. This characteristic underlies its potent antioxidant action and enzyme cofactor properties, and supports a complex thiol-exchange system, which hierarchically regulates cell activity.

3. Synthesis of hydrophobic QDs

The synthesis of the most frequently used semiconducting colloidal QDs, consisted of metal chalcogenides (sulphides, selenides and tellurides), is based either on the usage of organometallic precursors (e.g. dimethylcadmium, diethylzinc), metallic oxide (e.g. CdO, ZnO) or metallic salts of inorganic and organic acids (e.g. zinc stearate, cadmium acetate, cadmium nitrate (Bae et al. 2009)). The sources of chalcogenide anion are usually pure chalcogen elements (e.g. S, Se, Te). Whatever precursor is used, the resulted QDs are hydrophobic, but their quantum yields (QY) are higher (in the range of 20–60 %) compared to the QDs prepared by aqueous synthesis route (below 30 %). However, the trend is to avoid the usage of organometallic precursors, because they are less environmentally benign compared to other ones, which are more preferable (Mekis et al. 2003).

The most common approach to the synthesis of the colloidal hydrophobic QDs is the controlled nucleation and growth of particles in a solution of organometallic/chalcogen precursors containing the metal and the anion sources. The method lies in rapid injection of a solution of chemical reagents into a hot and vigorously stirred coordinating organic solvent (typically trioctylphosphine oxide (TOPO) or trioctylphosphine (TOP)) that can coordinate with the surface of the precipitated QDs particles (Talapin et al. 2010). Consequently, a large number of nucleation centres are initially formed at about 300 °C. The coordinating ligands in the hot solvents prevent or limit subsequent crystal growth (aggregation) via Ostwald ripening process (small crystals, which are more soluble than the large ones, dissolve and reprecipitate onto larger particles), which typically occurs at temperatures in the range of 250–300 °C (Merkoci 2009). Further improvement of the resulting size distribution of the QDs particles can be achieved through selective preparation (Mićić & Nozik 2002). Because these QDs are insoluble in aqueous solution and soluble in nonpolar solvents only, further functionalization is required to achieve their solubilization. However, this inconveniency is compensated with higher QY of these QDs as mentioned previously.

3.1 Solubilization of hydrophobic QDs

Solubilization of QDs is essential for many biological and biomedical applications and presents a significant challenge in this field. Transformation process is complicated and involves multiple steps. Different QDs solubilization strategies have been discovered over the past few years. Non-water soluble QDs can be grown easily in hydrophobic organic solvents, but the solubilization requires sophisticated surface chemistry alteration. Current methods for solubilization without affecting key properties are mostly based on exchange of the original hydrophobic surfactant layer (TOP/TOPO) capping the QDs with hydrophilic one or the addition of a second layer (Jamieson et al. 2007). However, in most cases, the surface exchange results in not only broadening of the size distribution but also in reductions of QY from 80% in the organic phase to about 40% in aqueous solution (Tian et al. 2009).

The first technique involves ligand exchange (sometimes called cap exchange). The native hydrophobic ligands are replaced by bifunctional ligands of surface anchoring thiol-containing molecules (see Fig. 2.) (usually a thiol, e.g. sodium thioglycolate) or more sophisticated ones (based on e.g. carboxylic or amino groups) such as oligomeric phosphines, dendrons and peptides to bind to the QDs surface and hydrophilic end groups (e.g. hydroxyl and carboxyl) to render water solubility. The second strategy employs polymerized silica shells functionalized with polar groups using a silica precursor during the polycondensation to insulate the hydrophobic QDs. While nearly all carboxy-terminated ligands limit QDs dispersion to basic pH, silica shell encapsulation provides stability over much broader pH range. The third method maintains native ligands on the QDs and uses variants of amphiphilic diblock and triblock copolymers and phospholipids to tightly interleave the alkylphosphine ligands through hydrophobic interactions (Michalet et al. 2005; Xing et al. 2009). Aside from rendering water solubility, these surface ligands play a critical role in insulating, passivating and protecting the QD surface from deterioration in biological media (Cai et al. 2007).

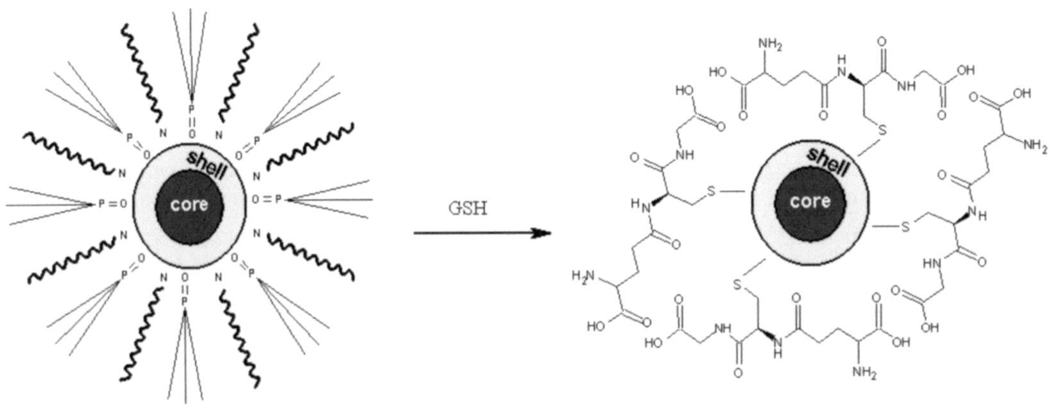

Fig. 2. Schematic representation of water soluble GSH-QDs preparation

An interesting work dealing with synthesis of hydrophobic QDs using chalcogen and metal oxide precursors and their following solubilisation with GSH was recently published by Jin et al. (Jin et al. 2008). The authors prepared highly fluorescent, water-soluble GSH-coated CdSeTe/CdS QDs emitting in near-infrared region (maximum emission at 800 nm) and tested them as optical contrast agents for *in vivo* fluorescence imaging. NIR emitting QDs are very suitable for *in vivo* imaging mainly due to low scattering and the absorption of NIR light in tissues. The preparation is based on surface modification of hydrophobic CdSeTe/CdS (core/shell) QDs with GSH in tetrahydrofuran-water solution. GSH is added in relatively high concentration of 30 mg for 1 ml of solution and its excess is finally removed by dialysis. The resulting GSH-QDs were stocked in PBS (pH = 7.4) and exhibited the QY of 22%.

Similarly, highly luminescent CdSe/ZnS QDs were synthesized by Gill and colleagues, who used GSH-capped QDs, which were further functionalized with fluorescein isothiocyanate-modified avidin (Gill et al. 2008). The resulting avidin-capped QDs were used in all ratiometric analyses of H_2O_2 and their fluorescence QY was about 20 %.

Tortiglione et al. prepared GSH-capped CdSe/ZnS QDs in three steps (Tortiglione et al. 2007). At first, they synthesized TOP/TOPO-capped CdSe/ZnS core/shell QDs via the pyrolysis of precursors, trioctylphosphine selenide and organometallic dimethylcadmium, in a coordinating solvent. Diethylzinc and hexamethyldisilathiane were used as Zn and S precursors, respectively in the formation of ZnS shell around CdSe core. Due to their hydrophobic properties, CdSe/ZnS QDs were subsequently transferred into aqueous solution by standard procedure of wrapping up them in an amphiphilic polymer shell (diamino-PEG 897). Finally, the PEG-QDs were modified with GSH via formation of an amide bond with free amino groups of the diamino-PEG. These functionalized fluorescent probes can be used for staining fresh water invertebrates (e.g. *Hydra vulgaris*).. GSH is known to promote Hydra feeding response by inducing mouth opening.

4. Aqueous synthesis of GSH coated QDs

The second and more utilized way is the aqueous synthesis, producing QDs with excellent water solubility, biological compatibility, and stability (usually more than two months). Compared with organic phase synthesis, aqueous synthesis exhibits good reproducibility, low toxicity, and it is inexpensive. Basically, the fabrication process of water-soluble QDs takes place in reflux condenser (usually in a three-necked flask equipped with this reflux condenser). Nevertheless, this procedure in water phase needs a very long reaction time ranging from several hours to several days. Recently, new strategies employing microwave-assisted (MW) synthesis, which seems to be faster compared to the reflux one, were published as well (see below).

The other disadvantages of QDs synthesized through aqueous route are the wider FWHM (the full width at half maximum) and lower QY which can attribute to defects and traps on the surface of nanocrystals (Y.-F. Liu & J.-S. Yu 2009). These defects can be eliminated by the selection of capping agents. The process of functionalization involves ligand exchange with thioalkyl acids such as thioglycolic acid (TGA) (Xu et al. 2008), mercaptoacetic acid (MAA) (Abd El sadek et al. 2011), mercaptopropionic acid (MPA) (Cui et al. 2007), mercaptoundecanoic acid (MUA) (Aldeek et al. 2008), mercaptosuccinic acid (MSA) (Huang et al. 2007) or reduced GSH.

From these ligands, GSH seems to be a very perspective molecule, since it provides an additional functionality to the QDs due to its key function in detoxification of heavy metals (cadmium, lead) in organism (Ali et al. 2007).

GSH is not only an important water-phase antioxidant and essential cofactor for antioxidant enzymes, but it also plays roles in catalysis, metabolism, signal transduction, and gene expression. Thus, GSH QDs as biological probe should be more biocompatible than other thiol-capping ligands. Concerning the application, GSH QDs can be used for easy determination of heavy metals regarding the fact, that the fluorescence is considerably quenched at the presence of heavy metals. Similarly, GSH QDs exhibit high sensitivity to H_2O_2 produced from the glucose oxidase catalysing oxidation of glucose and therefore glucose can be sensitively detected by the quenching of the GSH QDs florescence (Saran et al. 2011; J. Yuan et al. 2009).

4.1 QDs synthesis in reflux condenser

This synthesis route usually consists in reaction of heavy metal (Zn, Cd, ...) precursor with chalcogen precursors. Ordinarily used precursors of heavy metals easily dissolving in water are acetates, nitrates or chlorides. The chalcogen precursors can be either commercial solid powders (e.g. Na_2TeO_3 in the case of CdTe QDs) or freshly prepared before using in reaction procedure, e.g. H_2Te (preparation by adding sulphuric acid dropwise to the aluminium telluride (Al_2Te_3) (Zheng et al. 2007a)) or NaHTe (forming by reaction of sodium borohydride ($NaBH_4$) with Te powder (He et al. 2006; Zhang et al. 2003)) in the case of CdTe QDs. However, NaHTe and H_2Te are unstable compounds under ambient conditions; therefore the synthesis of CdTe QDs generally has to be performed in inert reaction systems (see Fig. 3.). Since Na_2TeO_3 is air-stable, all of operations can be performed in the air, avoiding the need for an inert atmosphere. The synthetic pathway is thus free of complicated vacuum manipulations and environmentally friendly.

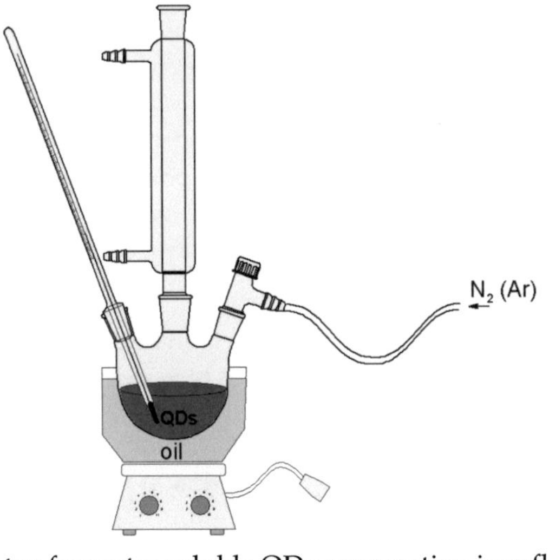

Fig. 3. Schema of apparatus for water soluble QDs preparation in reflux condenser

4.1.1 CdTe QDs capped with GSH

Xue et al. synthesized GSH-capped CdTe QDs by mixing the solutions of cadmium acetate and GSH and following injection of NaHTe solution under argon atmosphere and heating (Xue et al. 2011). After refluxing, QDs were precipitated with an equivalent amount of 2-propanol, followed by resuspension in a minimal amount of ultrapure water. Excess salts were removed by repeating this procedure three times, and the purified QDs were dried overnight at room temperature in vacuum. These GSH-QDs showed excellent photostability and possessed high QY (42 %) without any post-treatment. The authors conjugated the QDs with folic acid and studied how these labelled QDs can specifically target folic acid receptor on the surface of human hepatoma and human ovarian cancer cell to demonstrate their potentially application as biolabels.

Another GSH-functionalized QDS, namely CdTe and CdZnSe, were prepared by Ali et al. (Ali et al. 2007). The first mentioned were synthesized from H_2Te and $CdCl_2$, while in the second case NaHSe, $ZnCl_2$, H_2Se were used. Both types of GSH-capped QDs were coupled with a high-throughput detection system, to provide quick and ultrasensitive Pb^{2+} detection without the need of additional electronic devices. The mechanism is based on selective reduction of GSH-capped QDs in the presence of Pb^{2+} which results in fluorescence quenching that can be attributed to the stronger binding between heavy metal ions and the surface of GSH capping layer.

Also Goncalves and colleagues employed the simple experimental procedure for GSH-capped CdTe QDs fabrication and investigated the fluorescence intensity quenching in the presence of Pb^{2+} ions (Goncalves et al. 2009). Briefly, they mixed $CdCl_2$ and GSH aqueous solutions with freshly prepared NaHTe solution and the mixture was refluxed up to 8 h.

The same reactants for the synthesis of GSH-capped CdTe QDs were used by Cao M. et al. (Cao et al. 2009) and Dong et al. (Dong et al. 2010a). Cao and co-authors studied QDs interactions (fluorescence quenching) with heme-containing proteins and they found their optical fluorescence probes can be used for the selective determination of cytochrome c under optimal pH value. While Dong et al. used their GSH-CdTe QDs as fluorescent labels to link bovine serum albumin (BSA) and rat anti-mouse CD4, which was expressed on mouse T-lymphocyte and mouse spleen tissue. The authors demonstrated that CdTe QDs-based probe exhibited much better photostability and fluorescence intensity than one of the most common fluorophores, fluorescein isothiocyanate (FITC), showing a good application potential in the immuno-labeling of cells and tissues.

Wang and colleagues reported on the preparation of three kinds of water-soluble QDs, MAA-capped CdTe QDs, MAA-capped CdTe/ZnS and GSH-capped CdTe QDs, and compared the change of their fluorescence intensity (quenching) in the presence of As (III)(Wang et al. 2011). Arsenic (III) has a high affinity to reduced GSH to form $As(SG)_3$ thus the fluorescence of GSH coated QDs is reduced significantly in the presence of As (III). MAA-capped CdTe QDs were prepared through reaction of $CdCl_2$ and MAA with subsequent injection of freshly prepared NaHTe solution under vigorous magnetic stirring. Then the precursor solution was heated and refluxed under N_2 protection for 60 min. Finally, cold ethanol was added and MAA-CdTe QDs were precipitated out by centrifugation. A similar procedure was used for GSH-capped CdTe QDs synthesis with only one difference: the precipitation process was repeated for three times in order to eliminate free GSH ligands and salts in the GSH-CdTe QDs colloids. MAA-capped CdTe/ZnS QDs were prepared also similarly. When the CdTe precursor was refluxed for 30 min, $ZnCl_2$ and Na_2S were added slowly and simultaneously to form ZnS shell. After 30 min, the products were separated by the addition of cold ethanol and centrifugation.

Different thiol ligands, including TGA, L-cysteine (L-Cys) and GSH for capping CdTe QDs were also tested by Li Z. et al. (Z. Li et al. 2010). The starting materials were identical as in previous mentioned studies, i.e. NaHTe and $CdCl_2$. The luminescent properties of CdTe QDs with different stabilizing agents were studied by using fluorescence spectra, which showed that CdTe QDs with longer emission wavelength (680 nm) can be synthesized more easily when L-Cys or GSH is chosen as stabilizing agents. Moreover, the authors found that the cytotoxicity of TGA-QDs is higher than that of L-Cys- and GSH-CdTe. Ma et al. also prepared CdTe QDs modified with these three thiol-complex, namely TGA, L-cys and GSH and investigated the interactions of prepared QDs with BSA using spectroscopic methods (UV-VIS, IR and fluorescence spectrometry) (Ma et al. 2010).

Tian et al. (Tian et al. 2009) used for the first time GSH and TGA together to enhance stability of water soluble CdTe QDs prepared using NaHTe and $CdCl_2$. The author prepared different-sized CdTe QDs with controllable photoluminescence wavelengths from 500 to 610 nm within 5 h at temperature of 100 °C. When the molar ratio of GSH to TGA is 1:1, QY of the yellow-emitting CdTe (emission maximum at 550 nm) reached 63 % without any post-treatment. The synthesized CdTe QDs possess free carboxyl and amino groups, which were successfully conjugated with insulin for delivery to cells, demonstrating that they can be easily bound bimolecularly and have potentially broad applications as bioprobes.

Yuan et al. replaced NaHTe with more convenient Na_2TeO_3 for preparation of CdTe QDs, namely they used $CdCl_2$ and Na_2TeO_3, which were subsequently mixed with MSA or GSH as capping agent (J. Yuan et al. 2009). The prepared QDs were tested for glucose detection by monitoring QDs photoluminescence quenching as consequence of H_2O_2 presence and acidic changes produced by glucose oxidase catalysing glucose oxidation, respectively. The authors found that the sensitivity of QDs to H_2O_2 depends on QDs size: smaller size presented higher sensitivity. The quenching effect of H_2O_2 on GSH-capped QDs was more than two times more intensive than that on MSA-capped QDs.

4.1.2 CdSe QDs capped with GSH

Compared to CdTe QDs, GSH-capped CdSe QDs are much readily prepared. Jing et al. synthesized TGA-capped CdSe QDs using $CdCl_2$ and Na_2SeO_3, and they used these QDs for hydroxyl radical electrochemiluminescence sensing of the scavengers (Jiang & Ju 2007).

The research group of Dong, mentioned in synthesis of CdTe QDs, also prepared two kinds of highly fluorescent GSH-capped CdSe/CdS core-shell QDs emitting green and orange fluorescence at 350 nm excitation by an aqueous approach (Dong et al. 2010b). The authors used these QDs as fluorescent labels to link mouse anti-human CD3 which was expressed on human T-lymphocyte. Compared to CdSe QDs, they found a remarkable enhancement in the emission intensity and a red shift of emission wavelength for both types of core-shell CdSe/CdS QDs. They demonstrated that the fluorescent CdSe/CdS QDs exhibited much better photostability and brighter fluorescence than FITC.

4.1.3 CdS QDs capped with GSH

Also thiol-capped CdS QDs are less studied in comparison with CdTe QDs. MPA belongs to the most tested thiol ligands for capping these QDS (Huang et al. 2008). Liang et al. synthesized GSH-capped CdS QDs in aqueous solutions from $CdCl_2$ and CH_3CSNH_2 (thioacetamide) at room temperature (Liang et al. 2010). In this synthesis procedure, GSH was added in the final step into previously prepared CdS QDs solution. The obtained GSH coated QDs were tested as fluorescence probes to determine of Hg^{2+} with high sensitivity and selectivity. Under optimal conditions, the quenched fluorescence intensity increased linearly with the concentration of Hg^{2+}.

Merkoci et al. employed another preparation process: GSH and $CdCl_2$ were first dissolved in water with subsequent addition of TMAH (tetramethylammoniumhydroxide) and ethanol. After degassing, HMDST (hexamethyldisilathiane) was quickly added as sulphide precursor, giving a clear (slightly yellow) colloidal solution of water soluble CdS QDs modified with GSH (Merkoci et al. 2007). The authors used these QDs as a model compound in a direct electrochemical detection of CdS QDs or other similar QDs, based on the square-wave voltammetry of CdS QDs suspension dropped onto the surface of a screen printed electrode. This detection method is simple and low cost compared to optical methods and it will be interesting for bioanalytical assays, where CdS QDs can be used as electrochemical tracers, mainly in fast screening as well as in field analysis.

Thangadurai and colleagues investigated 5 organic thiols as suitable capping agent for CdS QDs (diameter of 2–3.3 nm), namely 1,4-dithiothreitol (DTT), 2-mercaptoethanol , L-Cys, methionine and GSH (Thangadurai et al. 2008). The QDs were prepared by a wet chemical method from $Cd(NO_3)_2$ and Na_2S. Briefly, the process started with addition of capping agent aqueous solution to the solution of $Cd(NO_3)_2$ and stirred for 12 h at room temperature and under dry N_2 atmosphere. In the second step, Na_2S solution was added dropwise and stirred for another 12 h. The CdS prepared with and without coating appeared greenish yellow and dark orange, respectively. The authors revealed the CdS QDs being in cubic phase. According to FT-IR studies, they suggested two different bonding mechanisms of the capping agents with the CdS. DTT was found to be the best capping agent for CdS from all tested thiols because of lower grain size in cubic phase and good fluorescence properties with efficient quenching of the surface traps.

Jiang et al. prepared GSH-capped aqueous CdS QDs with strong photoluminescence (QY of 36 %) using $CdCl_2$ and Na_2S by typical procedure (Jiang et al. 2007). The excitation spectrum was broad ranging from 200 to 480 nm. These QDs were conjugated with BSA and tested as fluorescence probes. The results demonstrated that the fluorescence of CdS QDs can be enhanced by BSA depending on BSA concentration.

4.1.4 Zn-based QDs capped with GSH

Generally, the QDs fluorescent colour can be tuned by changing their size which depends mainly on reaction time. There is also another option how to tune the colour of QDs emission without changing the QDs size using alloyed QDs, which is the most frequently used approach for Zn-based QDs. Alloyed QDs are traditionally fabricated in two step synthesis route, for example by incorporation of Cd^{2+} into very small ZnSe seeds (Zheng et al. 2007b). Subsequent stabilization of these QDs is usually ensured with thiol compounds. Cao et al. prepared water-soluble violet-green emitting core/shell $Zn_{1-x}Cd_xSe$/ZnS QDs using N-acetyl-l-cysteine (NAC) as a stabilizer (Cao et al. 2010). ZnS shell provided reduction of $Zn_{1-x}Cd_xSe$ core cytotoxicity and increase of QY up to 30 %, while NAC resulted in excellent biocompatibility of these QDs.

Liu and colleagues synthesized alloyed $Zn_xHg_{1-x}Se$ QDs capped with GSH in one step process by reacting a mixture of $Zn(ClO_4)_2$, $Hg(ClO_4)_2$ and GSH with freshly prepared NaHSe (Liu et al. 2009). The fluorescent color of the alloyed QDs can be easily tuned in the range of 548–621 nm by varying the Zn^{2+}:Hg^{2+} molar ratio, reaction pH, intrinsic Zn^{2+} and Hg^{2+} reactivity toward NaHSe, and the concentration of NaHSe. These GSH-capped $Zn_{0.96}Hg_{0.04}Se$ QDs possessed high QY (78 %) and were applied for sensing Cu^{2+}. Ying et al. synthesized another type of alloyed QDs, namely GSH-capped $Zn_{1-x}Cd_xSe$ QDs with tunable fluorescence emissions (360–700 nm) and QY up to 50 %(Ying et al. 2008). Lesnyak and colleagues demonstrated a facile one-step aqueous synthesis of blue-emitting GSH-capped $ZnSe_{1-x}Te_x$ QDs with QY up to 20 % (Lesnyak et al. 2010). Li et al. prepared GSH-capped alloyed $Cd_xZn_{1-x}Te$ QDs through a one-step aqueous route (W.W. Li et al. 2010). These QDs with high QY up to 75 % possessed broadened band gap, hardened lattice structure and lower defect densities. Their emission wavelength can be tuned from 470 to 610 nm. The authors suggested the usage of such QDs as promising optical probes in bio-applications or in detection of heavy metal ions (e.g. Pb^{2+}, Hg^{2+}).

Deng et al. examined two other thiol ligands beside GSH, MPA and TGA, for stabilization of ZnSe and $Zn_xCd_{1-x}Se$ QDs synthesized by water-based route (Deng et al. 2009). A typical synthetic procedure for ZnSe QDs started with mixing $Zn(NO_3)_2$, thiol molecule and N_2H_4 (hydrazine), which was used to maintain oxygen-free conditions, allowing the reaction vessel to be open to air. In the next step, freshly prepared NaHSe solution was added to the flask with vigorous stirring and the pH was adjusted to 11 using 1 M NaOH. The mixture was refluxed at temperature close to 100 °C which resulted in light blue solution as ZnSe QDs grew. The prepared QDs possessed tunable and narrow photoluminescence (PL) peaks ranging from 350 to 490 nm. The authors found that MPA capping agent gave rise to smaller ZnSe QDs with a high density of surface defects, while TGA and GSH produced larger ZnSe QDs with lower surface defect densities. According to absorption spectra, the growth was more uniform and better controlled with linear two-carbon TGA (QDs size of 2.5 nm) than with GSH, which is branched bifunctional molecule. Concerning $Zn_xCd_{1-x}Se$ QDs, the preparation was performed in a reducing atmosphere by addition of Cd-thiol complex directly to ZnSe QDs solution. The PL peaks changed from 400 to 490 nm by changing the Zn to Cd ratio.

Fang et al. fabricated water-dispersible GSH-capped ZnSe/ZnS core/shell QDs with high QY up to 65 % (Fang et al. 2009). In the first step, GSH-capped ZnSe core was synthesized by mixing zinc acetate with GSH solution. The pH of solution was adjusted to 11.5 by addition of 2 M NaOH. Subsequently, fresh NaHSe solution was added at room temperature. The system was heated to 90 °C under N_2 atmosphere for 1 h which resulted in formation of ZnSe core with an average size of 2.7 nm. In the second step, ZnS shell was created in reaction of as-prepared ZnSe core with shell precursor compounds (zinc acetate as zinc resource and thiourea as sulphur resource) at 90 °C. In comparison to the plain ZnSe QDs, both the QY and the stability against UV irradiation and chemical oxidation of ZnSe/ZnS core/shell QDs have been greatly improved.

4.2 Microwave irradiation synthesis

As mentioned above, long reaction times in aqueous phase often result in a large number of surface defects on synthesized QDs with low photoluminescence QY. Hydrothermal and microwave (MW) irradiation methods can replace traditional reflux methods and provide high-quality QDs in shorter time (Zhu et al. 2002). Especially, MW synthesis is advantageous due to rapid homogeneous heating realized through the penetration of microwaves. Compared to conventional thermal treatment, this way of heating allows the elimination of defects on QDs surface and produces uniform products with higher QY (Duan et al. 2009). The sizes of QDs can be easily tuned by varying the heating times. The QDs growth stops when the MW irradiation system is off and product is cooled down.

From chemical point of view, the most frequent types of QDs synthesized using microwave irradiation are CdTe, CdSe, CdS, $Zn_{1-x}Cd_xSe$ and ZnSe. As usual, these QDs can be functionalized with various thiol ligands such as MPA, MSA (Kanwal et al. 2010), TGA, 1-butanethiol, 2-mercaptoethanol (Majumder et al. 2010) or GSH (Qian et al. 2006). However, thiol ligands can be also used as sulphur source in one-step MW synthesis of QDs. Qian et al. reported on a seed-mediated and rapid synthesis of CdSe/CdS QDs using MPA, which was decomposed during MW irradiation releasing S^{2-} anions at temperature of 100 °C (Qian et al. 2005). In this step, only CdSe monomers were nucleated and grown by the reaction of NaHSe and cadmium chloride. The initial core was rich in Se due to the faster reaction of Se with Cd^{2+} compared to S. The amount of released S^{2-} anions increased, when the temperature rose to 140 °C which resulted in formation of alloyed CdSeS shell on the surface of CdSe nanocrystals. The resulted QDs showed the quantum yield up to 25 %.

Traditionally, GSH was used as thiol-capping agent for CdTe QDs in the work of other research group (Qian et al. 2006). Highly luminescent, water-soluble, and biocompatible CdTe QDs were synthesized in one-pot through reaction of Cd^{2+}–GSH complex (using cadmium chloride as Cd source) with freshly prepared NaHTe in a sealed vessel under MW irradiation at 130 °C in less than 30 min. The prepared nanocrystals possessed excellent optical properties and QY above 60 %. It is worth to note, that CdTe nanocrystals were tightly capped by Cd^{2+}–GSH at a lower pH value (compared to other thiol ligands, e.g pH 11.2 in the case of MSA (Kanwal et al. 2010)), which inhibited the growth of the nanocrystals. With the decrease of pH value, the growth rate slows dramatically.

A similar approach for one-step synthesis of GSH-capped ZnSe QDs in aqueous media was employed in the work of Huang et al. (Huang & Han 2010). The process was based on the reaction of air-stable Na_2SeO_3 with aqueous solution consisted of zinc nitrate and GSH. Then $NaBH_4$ as reduction agent was added into the above mentioned solution with stirring. The pH was set to value of 10 by the addition of NaOH. The mixture was then refluxed at 100 °C for 60 min under MW irradiation (300 W). The obtained QDs (2–3 nm), performed strong band-edge luminescence (QY reached 18%).

4.3 Microemulsion synthesis

This fabrication route is widely used for QDs coated with thiol ligands, however, according to our best knowledge, only one publication deals with GSH as coating material. Saran and colleagues employed this technique for fabrication of various core-shell QDs, namely CdSe/CdS, CdSe/ZnS and CdS/ZnS) (Saran et al. 2011). Following, the authors tested three ligands: mercaptoacetic acid, mercaptopropionic acid and GSH to find the optimal capping agent for glucose monitoring (biosensing) in human blood, which is essential for diagnosis of diabetes. These optical biosensors, based on QDs conjugated with glucose oxidase using carbodiimide bioconjugation method, work on the phenomenon of fluorescence quenching with simultaneous release of H_2O_2, which is detected then.

The microemulsion synthesis method is a simple, inexpensive and highly reproducible method, which enables excellent control of nanoparticles size and shape (Saran & Bellare 2010). This control of particle size is achieved simply by varying water-to-surfactant molar ratio. Nevertheless, the microemulsion synthesis gives relatively low yield of product; even large amounts of surfactant and organic solvent are used compared to bulk aqueous precipitation. The key point of this procedure is extraction of the nanoparticles from microemulsion into aqueous phase and to maintain their structural and surface features. In order to reach feasible yields of nanoparticles, the higher concentration of precursors in microemulsion should be used, which leads to much larger particle density inside the reverse micelles.

Briefly, a typical microemulsion synthesis of CdSe QDs can be described as follows: Se powder is added to Na_2SO_3 solution under continuous nitrogen bubbling at higher temperature forming Na_2SeSO_3 (sodium selenosulfate). Subsequently, this precursor was mixed with reverse micelle system prepared by dissolving AOT (sodium bis (2-ethylhexyl) sulfosuccinate) in n-heptane. A similar microemulsion was prepared with $Cd(NO_3)_2$. Finally, these two microemulsions were vortex-mixed which leaded to formation of CdSe QDs inside the reverse micelles. In the second step, a shell of CdS was created by the addition of $(NH_4)_2S$ microemulsion under vortex-stirring. The last step consisted in core-shell QDs stabilization using thiol ligands aqueous solution, which is added to the solution of QDs. The process is accompanied with colour change of organic phase (initially orange-red) to translucent. This colour change indicated the complete transfer of thiol-capped QDs into the aqueous phase (Fig. 4.).

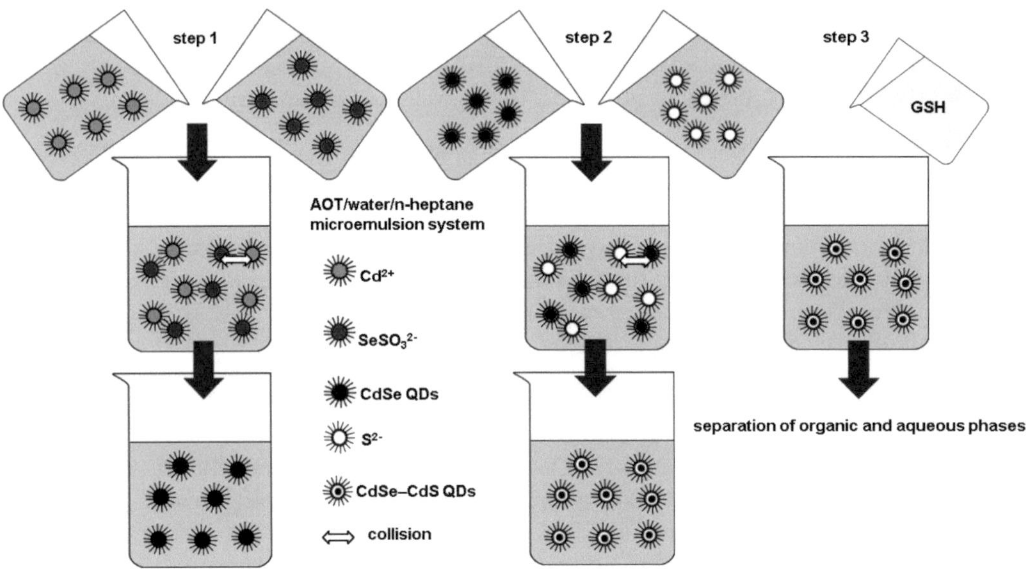

Fig. 4. Surface functionalization, recovery and stabilization of QDs from microemulsion into aqueous phase

5. Conclusion

Current issues solved in synthesis of highly luminescent QDs are their easy preparation, biocompatibility, stability and solubility in water. Up to now, the most frequently used approaches reported on the preparation of colloidal QDs are (1) synthesis of hydrophobic QDs with subsequent solubilization step, (2) direct aqueous synthesis or (3) two-phase synthesis. Compared with hydrophobic or two-phase approaches, aqueous synthesis is reagent-effective, less toxic and more reproducible. There is a variety of capping ligands used to provide solubility and biocompatibility of QDs in aqueous synthesis, mainly thiol organic compounds. Among them, GSH has gained the most attention due to its excellent properties and application in detection or sensing purposes. Our chapter describes the most commonly used techniques for preparation of various GSH-coated QDs based on heavy metal chalcogenides, namely CdTe, CdS, CdSe and alloyed or simple Zn-based QDs.

6. Acknowledgement

This work has been supported by Grant Agency of the Academy of Sciencies of the Czech Republic under the contract GAAV KAN208130801 (NANOSEMED) by Grant Agency of the Czech Republic under the contract GACR 102/10/P618 and by project CEITEC CZ.1.05/1.1.00/02.0068.

7. References

Abd El-sadek M.S.; Nooralden A.Y.; Babu S.M. & Palanisamy P.K. (2011). Influence of different stabilizers on the optical and nonlinear optical properties of CdTe nanoparticles. *Optics Communications*, Vol. 284, No. 12, pp. 2900-2904, ISS N 0030-4018

Aldeek F.; Balan L.; Lambert J. & Schneider R. (2008). The influence of capping thioalkyl acid on the growth and photoluminescence efficiency of CdTe and CdSe quantum dots. *Nanotechnology*, Vol. 19, No. 47, pp. ISS N 0957-4484

Ali E.M.; Zheng Y.G.; Yu H.H. & Ying J.Y. (2007). Ultrasensitive Pb2+ detection by glutathione-capped quantum dots. *Analytical Chemistry*, Vol. 79, No. 24, pp. 9452-9458, ISS N 0003-2700

Bae P.K.; Kim K.N.; Lee S.J.; Chang H.J.; Lee C.K. & Park J.K. (2009). The modification of quantum dot probes used for the targeted imaging of his-tagged fusion proteins. *Biomaterials*, Vol. 30, No. 5, pp. 836-842, ISS N 0142-9612

Byers R.J. & Hitchman E.R. (2011). Quantum Dots Brighten Biological Imaging. *Progress in Histochemistry and Cytochemistry*, Vol. 45, No. 4, pp. 201-237, ISS N 0079-6336

Cai W.B.; Hsu A.R.; Li Z.B. & Chen X.Y. (2007). Are quantum dots ready for in vivo imaging in human subjects? *Nanoscale Research Letters*, Vol. 2, No. 6, pp. 265-281, ISS N 1931-7573

Cao J.; Xue B.; Li H.; Deng D. & Gu Y. (2010). Facile synthesis of high-quality water-soluble N-acetyl-l-cysteine-capped Zn1−xCdxSe/ZnS core/shell quantum dots emitting in the violet–green spectral range. *Journal of Colloid and Interface Science*, Vol. 348, No. 2, pp. 369-376, ISS N 0021-9797

Cao M.; Cao C.; Liu M.G.; Wang P. & Zhu C.Q. (2009). Selective fluorometry of cytochrome c using glutathione-capped CdTe quantum dots in weakly basic medium. *Microchimica Acta*, Vol. 165, No. 3-4, pp. 341-346, ISS N 0026-3672

Chomoucka J.; Drbohlavova J.; Adam V.; Kizek R.; Hubalek J. & Ieee. (2009). *Synthesis of Glutathione-coated Quantum Dots*, Ieee, ISB N 978-1-4244-4260-7, New York

Chomoucka J.; Drbohlavova J.; Babula P.; Adam V.; Hubalek J.; Provaznik I. & Kizek R. (2010). Cell Toxicity and Preparation of Streptavidin-Modified Iron Nanoparticles and Glutathione-Modified Cadmium-Based Quantum Dots, *Proceedings of Eurosensor Xxiv Conference*, ISB N 1877-7058, Linz, 2010

Cui R.; Pan H.C.; Zhu J.J. & Chen H.Y. (2007). Versatile immunosensor using CdTe quantum dots as electrochemical and fluorescent labels. *Analytical Chemistry*, Vol. 79, No. 22, pp. 8494-8501, ISS N 0003-2700

Deng Z.T.; Lie F.L.; Shen S.Y.; Ghosh I.; Mansuripur M. & Muscat A.J. (2009). Water-Based Route to Ligand-Selective Synthesis of ZnSe and Cd-Doped ZnSe Quantum Dots with Tunable Ultraviolet A to Blue Photoluminescence. *Langmuir*, Vol. 25, No. 1, pp. 434-442, ISS N 0743-7463

Dong W.; Ge X.; Wang M. & Xu S.K. (2010a). Labeling of BSA and imaging of mouse T-lymphocyte as well as mouse spleen tissue by L-glutathione capped CdTe quantum dots. *Luminescence*, Vol. 25, No. 1, pp. 55-60, ISS N 1522-7235

Dong W.; Ge X.; Wang X.Y. & Xu S.K. (2010b). Preparation of GSH Capped CdSe/CdS Core-Shell QDs and Labeling of Human T-Lymphocyte. *Spectroscopy and Spectral Analysis*, Vol. 30, No. 1, pp. 118-122, ISS N 1000-0593

Drummen G.P.C. (2010). Quantum Dots-From Synthesis to Applications in Biomedicine and Life Sciences. *International Journal of Molecular Sciences*, Vol. 11, No. 1, pp. 154-163, ISS N 1422-0067

Duan J.L.; Song L.X. & Zhan J.H. (2009). One-Pot Synthesis of Highly Luminescent CdTe Quantum Dots by Microwave Irradiation Reduction and Their Hg(2+)-Sensitive Properties. *Nano Research*, Vol. 2, No. 1, pp. 61-68, ISS N 1998-0124

Fang Z.; Li Y.; Zhang H.; Zhong X.H. & Zhu L.Y. (2009). Facile Synthesis of Highly Luminescent UV-Blue-Emitting ZnSe/ZnS Core/Shell Nanocrystals in Aqueous Media. *Journal of Physical Chemistry C*, Vol. 113, No. 32, pp. 14145-14150, ISS N 1932-7447

Gill R.; Bahshi L.; Freeman R. & Willner I. (2008). Optical detection of glucose and acetylcholine esterase inhibitors by H2O2-sensitive CdSe/ZnS quantum dots. *Angewandte Chemie-International Edition*, Vol. 47, No. 9, pp. 1676-1679, ISS N 1433-7851

Goncalves H.; Mendonca C. & da Silva J. (2009). PARAFAC Analysis of the Quenching of EEM of Fluorescence of Glutathione Capped CdTe Quantum Dots by Pb(II). *Journal of Fluorescence*, Vol. 19, No. 1, pp. 141-149, ISS N 1053-0509

He Y.; Lu H.T.; Sai L.M.; Lai W.Y.; Fan Q.L.; Wang L.H. & Huang W. (2006). Microwave-assisted growth and characterization of water-dispersed CdTe/CdS core-shell nanocrystals with high photoluminescence. *Journal of Physical Chemistry B*, Vol. 110, No. 27, pp. 13370-13374, ISS N 1520-6106

Helmut S. (1999). Glutathione and its role in cellular functions. *Free Radical Biology and Medicine*, Vol. 27, No. 9-10, pp. 916-921, ISS N 0891-5849

Huang C.-P.; Li Y.-K. & Chen T.-M. (2007). A highly sensitive system for urea detection by using CdSe/ZnS core-shell quantum dots. *Biosensors and Bioelectronics*, Vol. 22, No. 8, pp. 1835-1838, ISS N 0956-5663

Huang C.B.; Wu C.L.; Lai J.P.; Li S.Y.; Zhen J.S. & Zhao Y.B. (2008). CdS quantum dots as fluorescence probes for the detection of selenite. *Analytical Letters*, Vol. 41, No. 11, pp. 2117-2132, ISS N 0003-2719

Huang L. & Han H.Y. (2010). One-step synthesis of water-soluble ZnSe quantum dots via microwave irradiation. *Materials Letters*, Vol. 64, No. 9, pp. 1099-1101, ISS N 0167-577X

Jamieson T.; Bakhshi R.; Petrova D.; Pocock R.; Imani M. & Seifalian A.M. (2007). Biological applications of quantum dots. *Biomaterials*, Vol. 28, No. 31, pp. 4717-4732, ISS N 0142-9612

Jiang C.; Xu S.K.; Yang D.Z.; Zhang F.H. & Wang W.X. (2007). Synthesis of glutathione-capped US quantum dots and preliminary studies on protein detection and cell fluorescence image. *Luminescence*, Vol. 22, No. 5, pp. 430-437, ISS N 1522-7235

Jiang H. & Ju H.X. (2007). Electrochemiluminescence sensors for scavengers of hydroxyl radical based on its annihilation in CdSe quantum dots film/peroxide system. *Analytical Chemistry*, Vol. 79, No. 17, pp. 6690-6696, ISS N 0003-2700

Jin T.; Fujii F.; Komai Y.; Seki J.; Seiyama A. & Yoshioka Y. (2008). Preparation and Characterization of Highly Fluorescent, Glutathione-coated Near Infrared Quantum Dots for in Vivo Fluorescence Imaging. *International Journal of Molecular Sciences*, Vol. 9, No. 10, pp. 2044-2061, ISS N 1422-0067

Kanwal S.; Traore Z.; Zhao C. & Su X. (2010). Enhancement effect of CdTe quantum dots–IgG bioconjugates on chemiluminescence of luminol–H2O2 system. *Journal of Luminescence*, Vol. 130, No. 10, pp. 1901-1906, ISS N 0022-2313

Lesnyak V.; Dubavik A.; Plotnikov A.; Gaponik N. & Eychmuller A. (2010). One-step aqueous synthesis of blue-emitting glutathione-capped ZnSe(1-x)Te(x) alloyed nanocrystals. *Chemical Communications*, Vol. 46, No. 6, pp. 886-888, ISS N 1359-7345

Li W.W.; Liu J.; Sun K.; Dou H.J. & Tao K. (2010). Highly fluorescent water soluble Cd(x)Zn(1-x)Te alloyed quantum dots prepared in aqueous solution: one-step synthesis and the alloy effect of Zn. *Journal of Materials Chemistry*, Vol. 20, No. 11, pp. 2133-2138, ISS N 0959-9428

Li Z.; Wang Y.X.; Zhang G.X. & Han Y.J. (2010). Luminescent Properties and Cytotoxicity of CdTe Quantum Dots with Different Stabilizing Agents. *Journal of Inorganic Materials*, Vol. 25, No. 5, pp. 495-499, ISS N 1000-324X

Liang A.N.; Wang L.; Chen H.Q.; Qian B.B.; Ling B. & Fu J. (2010). Synchronous fluorescence determination of mercury ion with glutathione-capped CdS nanoparticles as a fluorescence probe. *Talanta*, Vol. 81, No. 1-2, pp. 438-443, ISS N 0039-9140

Liu F.-C.; Chen Y.-M.; Lin J.-H. & Tseng W.-L. (2009). Synthesis of highly fluorescent glutathione-capped ZnxHg1−xSe quantum dot and its application for sensing copper ion. *Journal of Colloid and Interface Science*, Vol. 337, No. 2, pp. 414-419, ISS N 0021-9797

Liu Y.-F. & Yu J.-S. (2009). Selective synthesis of CdTe and high luminescence CdTe/CdS quantum dots: The effect of ligands. *Journal of Colloid and Interface Science*, Vol. 333, No. 2, pp. 690-698, ISS N 0021-9797

Liu Y.F. & Yu J.S. (2009). Selective synthesis of CdTe and high luminescence CdTe/CdS quantum dots: The effect of ligands. *Journal of Colloid and Interface Science*, Vol. 333, No. 2, pp. 690-698, ISS N 0021-9797

Ma J.J.; Liang J.G. & Han H.Y. (2010). Study on the Synchronous Interactions between Different Thiol-Capped CdTe Quantum Dots and BSA. *Spectroscopy and Spectral Analysis*, Vol. 30, No. 4, pp. 1039-1043, ISS N 1000-0593

Majumder M.; Karan S.; Chakraborty A.K. & Mallik B. (2010). Synthesis of thiol capped CdS nanocrystallites using microwave irradiation and studies on their steady state and time resolved photoluminescence. *Spectrochimica Acta Part A: Molecular and Biomolecular Spectroscopy*, Vol. 76, No. 2, pp. 115-121, ISS N 1386-1425

Mekis I.; Talapin D.V.; Kornowski A.; Haase M. & Weller H. (2003). One-pot synthesis of highly luminescent CdSe/CdS core-shell nanocrystals via organometallic and "greener" chemical approaches. *Journal of Physical Chemistry B*, Vol. 107, No. 30, pp. 7454-7462, ISS N 1520-6106

Merkoci A. (2009). *Biosensing using nanomaterials*, Wiley, ISB N 978-0-470-18309-0, New Jersey

Merkoci A.; Marcolino-Junior L.H.; Marin S.; Fatibello-Filho O. & Alegret S. (2007). Detection of cadmium sulphide nanoparticles by using screen-printed electrodes and a handheld device. *Nanotechnology*, Vol. 18, No. 3, pp. ISS N 0957-4484

Michalet X.; Pinaud F.F.; Bentolila L.A.; Tsay J.M.; Doose S.; Li J.J.; Sundaresan G.; Wu A.M.; Gambhir S.S. & Weiss S. (2005). Quantum dots for live cells, in vivo imaging, and diagnostics. *Science*, Vol. 307, No. 5709, pp. 538-544, ISS N 0036-8075

Mićić O.I. & Nozik A.J. (2002). *Chapter 5 - Colloidal quantum dots of III–V semiconductors*. In: *Nanostructured Materials and Nanotechnology*, Hari Singh N., pp. 183-205, Academic Press, ISB N 978-0-12-513920-5, San Diego

Milne L.; Nicotera P.; Orrenius S. & Burkitt M.J. (1993). Effects of Glutathione and Chelating Agents on Copper-Mediated DNA Oxidation: Pro-oxidant and Antioxidant Properties of Glutathione. *Archives of Biochemistry and Biophysics*, Vol. 304, No. 1, pp. 102-109, ISS N 0003-9861

Qian H.; Li L. & Ren J. (2005). One-step and rapid synthesis of high quality alloyed quantum dots (CdSe–CdS) in aqueous phase by microwave irradiation with controllable temperature. *Materials Research Bulletin*, Vol. 40, No. 10, pp. 1726-1736, ISS N 0025-5408

Qian H.F.; Dong C.Q.; Weng J.F. & Ren J.C. (2006). Facile one-pot synthesis of luminescent, water-soluble, and biocompatible glutathione-coated CdTe nanocrystals. *Small*, Vol. 2, No. 6, pp. 747-751, ISS N 1613-6810

Ryvolova M.; Chomoucka J.; Janu L.; Drbohlavova J.; Adam V.; Hubalek J. & Kizek R. (2011). Biotin-modified glutathione as a functionalized coating for bioconjugation of CdTe-based quantum dots. *Electrophoresis*, Vol. 32, No. 13, pp. 1619-1622, ISS N 0173-0835

Saran A.D. & Bellare J.R. (2010). Green engineering for large-scale synthesis of water-soluble and bio-taggable CdSe and CdSe–CdS quantum dots from microemulsion by double-capping. *Colloids and Surfaces A: Physicochemical and Engineering Aspects*, Vol. 369, No. 1-3, pp. 165-175, ISS N 0927-7757

Saran A.D.; Sadawana M.M.; Srivastava R. & Bellare J.R. (2011). An optimized quantum dot-ligand system for biosensing applications: Evaluation as a glucose biosensor.

Colloids and Surfaces A: Physicochemical and Engineering Aspects, Vol. 384, No. 1-3, pp. 393-400, ISS N 0927-7757

Talapin D.V.; Lee J.S.; Kovalenko M.V. & Shevchenko E.V. (2010). Prospects of Colloidal Nanocrystals for Electronic and Optoelectronic Applications. *Chemical Reviews*, Vol. 110, No. 1, pp. 389-458, ISS N 0009-2665

Thangadurai P.; Balaji S. & Manoharan P.T. (2008). Surface modification of CdS quantum dots using thiols - structural and photophysical studies. *Nanotechnology*, Vol. 19, No. 43, pp. ISS N 0957-4484

Tian J.; Liu R.; Zhao Y.; Xu Q. & Zhao S. (2009). Controllable synthesis and cell-imaging studies on CdTe quantum dots together capped by glutathione and thioglycolic acid. *Journal of Colloid and Interface Science*, Vol. 336, No. 2, pp. 504-509, ISS N 0021-9797

Tortiglione C.; Quarta A.; Tino A.; Manna L.; Cingolani R. & Pellegrino T. (2007). Synthesis and biological assay of GSH functionalized fluorescent quantum dots for staining Hydra vulgaris. *Bioconjugate Chemistry*, Vol. 18, No. 3, pp. 829-835, ISS N 1043-1802

Wang X.; Lv Y. & Hou X. (2011). A potential visual fluorescence probe for ultratrace arsenic (III) detection by using glutathione-capped CdTe quantum dots. *Talanta*, Vol. 84, No. 2, pp. 382-386, ISS N 0039-9140

Wang Y. & Chen L. (2011). Quantum dots, lighting up the research and development of nanomedicine. *Nanomedicine: Nanotechnology, Biology and Medicine*, Vol. 7, No. 4, pp. 385-402, ISS N 1549-9634

Xing Y.; Xia Z.Y. & Rao J.H. (2009). Semiconductor Quantum Dots for Biosensing and In Vivo Imaging. *Ieee Transactions on Nanobioscience*, Vol. 8, No. 1, pp. 4-12, ISS N 1536-1241

Xu W.B.; Wang Y.X.; Liang S.; Xu R.H.; Zhang G.X.; Xu F.H. & Yin D.Z. (2008). Optimized synthesis and fluorescence spectrum analysis of CdSe quantum dots. *Journal of Dispersion Science and Technology*, Vol. 29, No. 7, pp. 953-957, ISS N 0193-2691

Xue M.; Wang X.; Wang H. & Tang B. (2011). The preparation of glutathione-capped CdTe quantum dots and their use in imaging of cells. *Talanta*, Vol. 83, No. 5, pp. 1680-1686, ISS N 0039-9140

Ying J.Y.; Zheng Y.G. & Selvan S.T. (2008) *Synthesis and applications of quantum dots and magnetic quantum dots*. In: *Colloidal Quantum Dots for Biomedical Applications*, Osinski M., Jovin T.M., Yamamoto K., pp. 86602-86602, Spie-Int Soc Optical Engineering, ISB N 0277-786X 978-0-8194-7041-6, Bellingham

Yuan J.; Guo W.; Yin J. & Wang E. (2009). Glutathione-capped CdTe quantum dots for the sensitive detection of glucose. *Talanta*, Vol. 77, No. 5, pp. 1858-1863, ISS N 0039-9140

Yuan J.P.; Guo W.W.; Yin J.Y. & Wang E.K. (2009). Glutathione-capped CdTe quantum dots for the sensitive detection of glucose. *Talanta*, Vol. 77, No. 5, pp. 1858-1863, ISS N 0039-9140

Zhang H.; Zhou Z.; Yang B. & Gao M.Y. (2003). The influence of carboxyl groups on the photoluminescence of mercaptocarboxylic acid-stabilized CdTe nanoparticles. *Journal of Physical Chemistry B*, Vol. 107, No. 1, pp. 8-13, ISS N 1520-6106

Zheng Y.G.; Gao S.J. & Ying J.Y. (2007a). Synthesis and cell-imaging applications of glutathione-capped CdTe quantum dots. *Advanced Materials,* Vol. 19, No. 3, pp. 376-+, ISS N 0935-9648

Zheng Y.G.; Yang Z.C.; Li Y.Q. & Ying J.Y. (2008). From glutathione capping to a crosslinked, phytochelatin-like coating of quantum dots. *Advanced Materials,* Vol. 20, No. 18, pp. 3410-+, ISS N 0935-9648

Zheng Y.G.; Yang Z.C. & Ying J.Y. (2007b). Aqueous synthesis of glutathione-capped ZnSe and Zn1-xCdxSe alloyed quantum dots. *Advanced Materials,* Vol. 19, No. 11, pp. 1475-+, ISS N 0935-9648

Zhu J.-J.; Wang H.; Zhu J.-M. & Wang J. (2002). A rapid synthesis route for the preparation of CdS nanoribbons by microwave irradiation. *Materials Science and Engineering: B,* Vol. 94, No. 2-3, pp. 136-140, ISS N 0921-5107

2

Quantum Measurement and Sub-Band Tunneling in Double Quantum Dots

Héctor Cruz
Universidad de La Laguna
Spain

1. Introduction

Quantum dots have attracted significant interest in recent years. Quantum dots attractive candidates as the building blocks for a quantum computer due to their potential to readily scale. The number of electrons can be reduced down to one in a gate-defined quantum dot.

High frequency operations on quantum dot systems have been used to observe new phenomena such as coherent charge oscillations and elastic tunneling behavior. Observation of these phenomena is made possible by (*in situ*) control of the rate of tunneling Γ between the quantum dots.

Measurements with a noninvasive detector in a double quantum dot system (qubit) has been extensively realized (Astley et al., 2007). A group of electrons is placed in a double quantum dot, whereas the detector (a quantum point contact) is localized near one of the dots. The quantum point contact acts as a measuring device.

One remaining key question is the theoretical study of the tunneling dynamics after the observation in a double quantum dot system (Cruz, 2002). Electrons can be projected onto a well define quantum dot after the observation takes place, if we consider the two quantum dots highly isolated (Ferreira et al., 2010).

In addition, we know that if two electron subbands are occupied, the electrical properties can be strongly modified due to the carrier-carrier interaction between subbands (Shabami et al., 2010). In this work we shall extend the Coulomb effect analysis when two subbands are occupied in the quantum dots. Then, the tunneling process could be modified due to the using of two different wave functions for two electron groups that interact between each other.

2. Model

It has been found that there are two distinct energy bands within semiconductors. From experiments, it is found that the lower band is almost full of electrons and the conduct by the movement of the empty states. In a semiconductor, the upper band is almost devoid of electrons. It represents excited electron states promoted from localized covalent bonds into extended states. Such electrons contribute to the current flow. The energy difference between the two bands is known as the band gap. Effective masses of around $0.067m_0$ for an electron in the conduction bands and $0.6m_0$ for a hole in the valence band can be taken in GaAs.

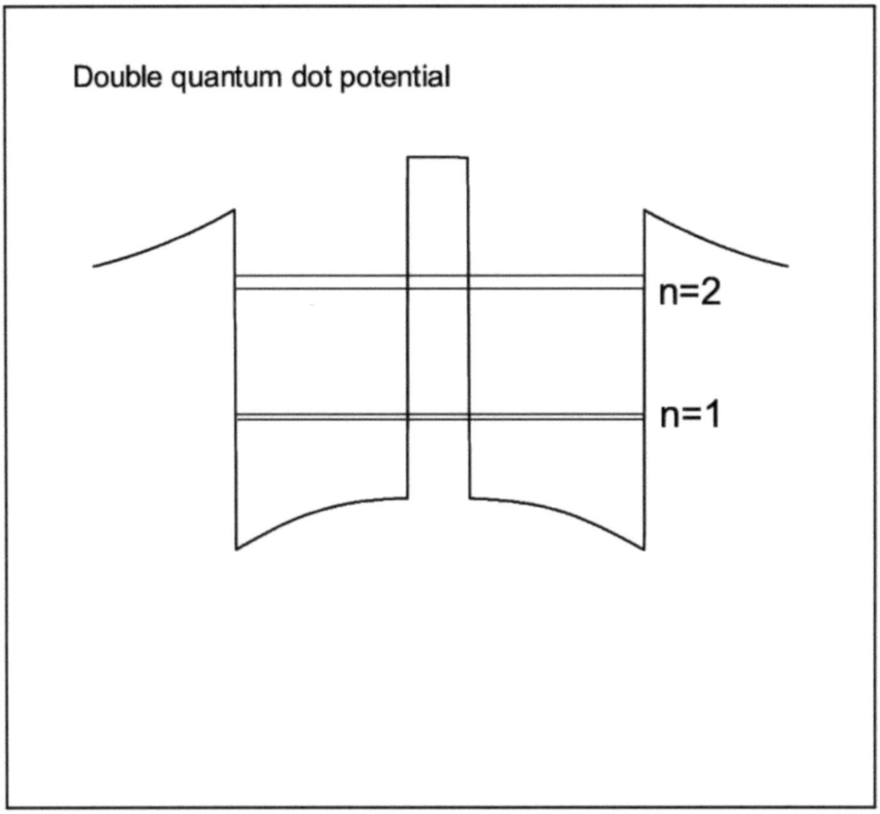

Fig. 1. A schematic illustration of the proposed experiment in the semiconductor quantum dot system. Double quantum dot system in absence of external bias.

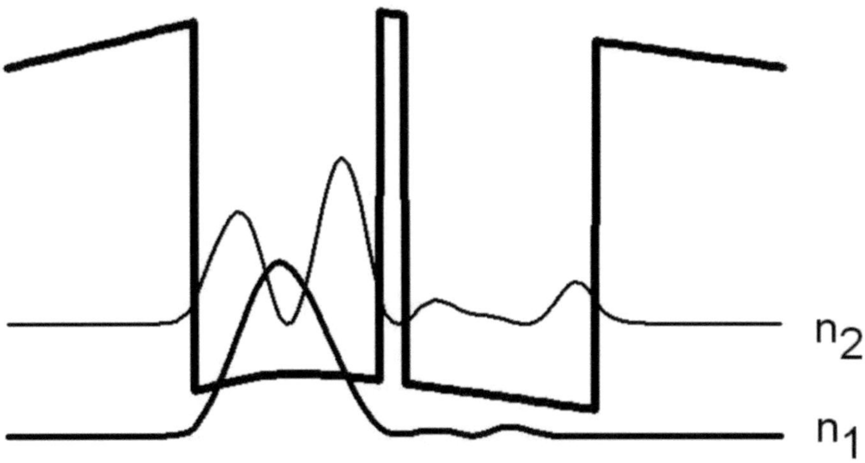

Fig. 2. Conduction band potential and carrier wave functions at $t = 0.1$ ps. We have taken an initial carrier density equal to $n_1 = 3.0 \times 10^{11}\text{cm}^{-2}$ and $n_2 = 0.0 \times 10^{11}\text{cm}^{-2}$.

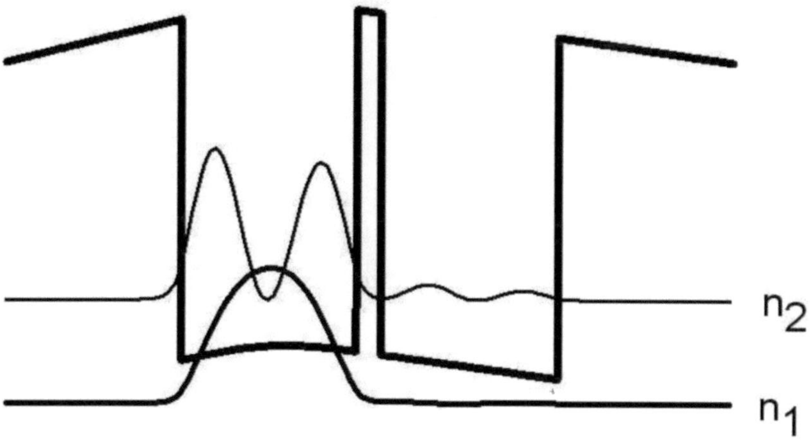

Fig. 3. Conduction band potential and carrier wave functions at $t = 0.2$ ps. We have taken an initial carrier density equal to $n_1 = 3.0 \times 10^{11}\text{cm}^{-2}$ and $n_2 = 0.0 \times 10^{11}\text{cm}^{-2}$.

The effective mass approximation is for a bulk crystal. The crystal is so large with respect to the scale of an electron wave function that is efectively infinite. In such a case, The Schrödinger equation has been found to be as follows:

$$-\frac{\hbar^2}{2m^*}\frac{\partial^2}{\partial z^2}\psi(z) = E\psi(z) \tag{1}$$

This equation is valid when two materials are placed adjacent to each other to form a heterojunction. The effective mass could be a function of the position and the band gaps of the materials can also be different. The discontinuity can be represented by a constant potential term. Thus the Schrödinger equation would be generalized to

$$-\frac{\hbar^2}{2m^*}\frac{\partial^2}{\partial z^2}\psi(z) + V(z)\psi(z) = E\psi(z) \tag{2}$$

The one dimensional potential $V(z)$ represents the band discontinuities at the heterojunction. The one dimensional potential is constructed from alternanting layers of dissimilar semiconductors, then the eletron or hole can move in the plane of the layers.

In this case, all the terms of the kinetic operator are required, and the Schrödinger equation would be as follows:

$$-\frac{\hbar^2}{2m^*}(\frac{\partial^2}{\partial x^2}\psi + \frac{\partial^2}{\partial y^2}\psi + \frac{\partial^2}{\partial z^2}\psi) + V(z)\psi = E\psi \tag{3}$$

As the potential can be written as a sum of independent functions, i.e.

$$V = V(x) + V(y) + V(z) \tag{4}$$

the eigenfunction of the system can be written as:

$$\psi(x, y, z) = \psi_x(x)\psi_y(y)\psi_z(z) \tag{5}$$

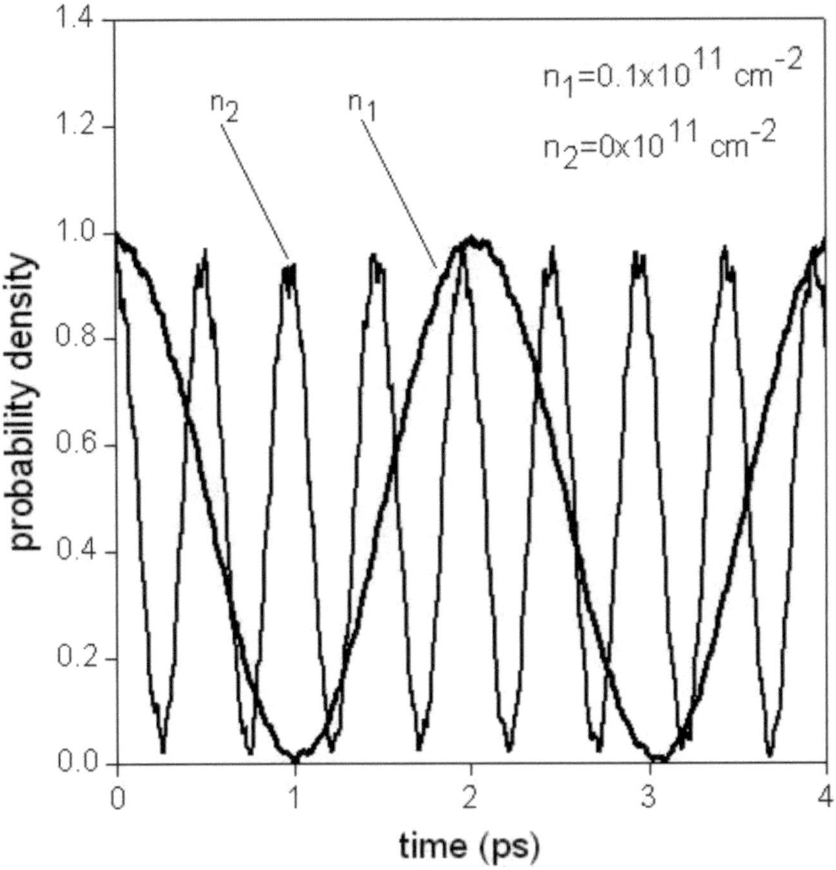

Fig. 4. Probability density in the left quantum well versus time at different carrier densities. $n_1 = 0.1 \times 10^{11}$ cm^{-2} and $n_2 = 0 \times 10^{11}$ cm^{-2}.

and using this in the above Schrödinger equation, then:

$$-\frac{\hbar^2}{2m^*}(\frac{\partial^2 \psi_x}{\partial x^2}\psi_y\psi_z + \frac{\partial^2 \psi_y}{\partial y^2}\psi_x\psi_z + \frac{\partial^2 \psi_z}{\partial z^2}\psi_x\psi_y) + V(z)\psi_x\psi_y\psi_z = E\psi_x\psi_y\psi_z \tag{6}$$

The last component is identical to a one-dimensional equation for a confining potential $V(z)$. The x and y components represent a moving particle and the wave function must reflect a current flow and have complex components. Then,

$$-\frac{\hbar^2}{2m^*}\frac{\partial^2}{\partial x^2}e^{ik_x x} = E_x e^{ik_x x} \tag{7}$$

and thus,

$$-\frac{\hbar^2}{2m^*}\frac{\partial^2}{\partial y^2}e^{ik_y y} = E_y e^{ik_y y} \tag{8}$$

where

$$\frac{\hbar^2}{2m^*}{k_x}^2 = E_x \tag{9}$$

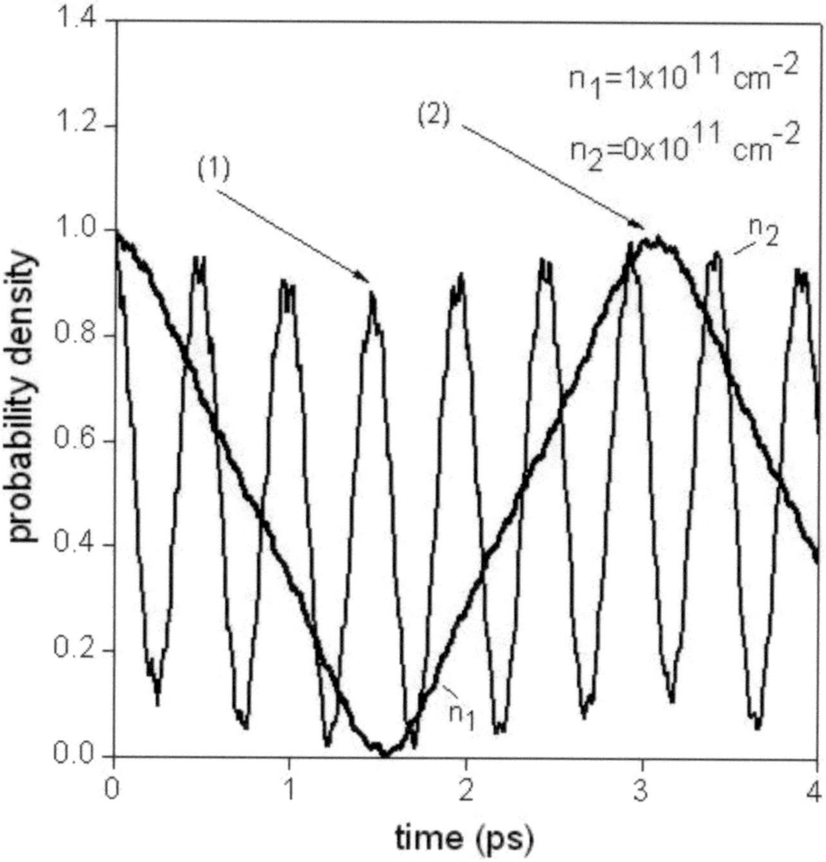

Fig. 5. Probability density in the left quantum well versus time at different carrier densities. $n_1 = 1 \times 10^{11}$ cm^{-2} and $n_2 = 0 \times 10^{11}$ cm^{-2}.

and

$$\frac{\hbar^2}{2m^*}{k_y}^2 = F_y \tag{10}$$

An infinity extent in the $x - y$ plane can be summarized as:

$$\psi_{x,y}(x,y) = \frac{1}{A}e^{i(k_x x + k_y y)} \tag{11}$$

and

$$E_{x,y} = \frac{\hbar^2 k_{x,y}^2}{2m^*} \tag{12}$$

Therefore, while solutions of the Schrödinger equation along the axis of the one-dimensional produce discrete states of energy E_z in the plane of a semiconductor quantum well, there is a continuous range of allowed energies.

In order to study the dynamics in the quantum well direction, we need to solve the time-dependent Schrödinger equation associated with an electron in a well potential for each

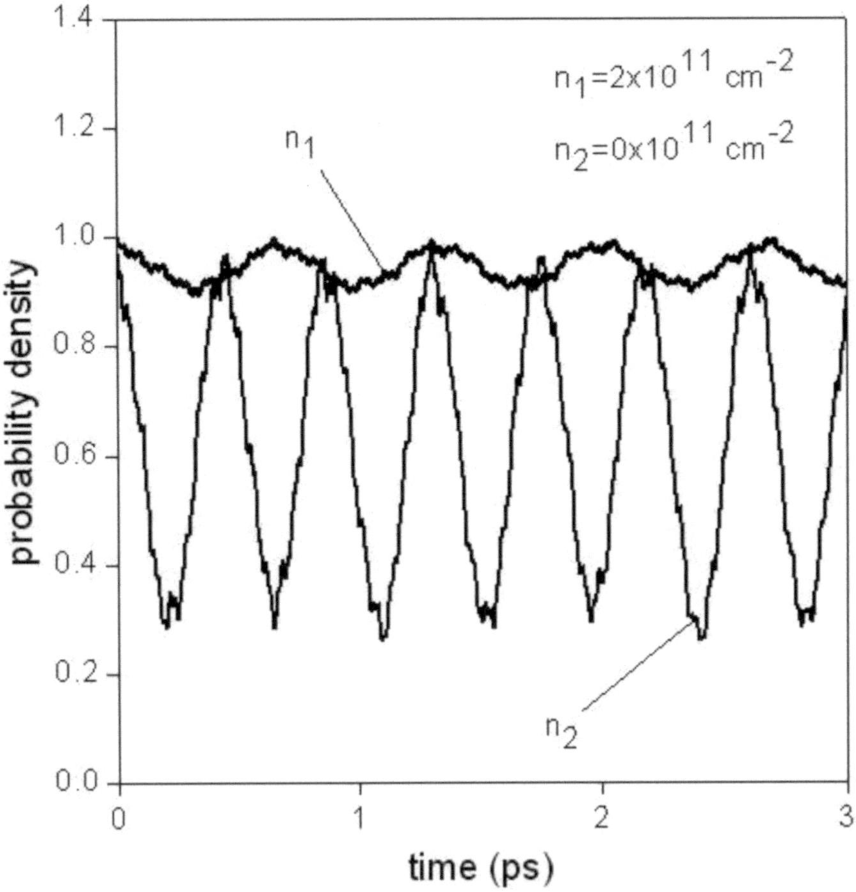

Fig. 6. Probability density in the left quantum well versus time at different carrier densities. $n_1 = 2 \times 10^{11}$ cm^{-2} and $n_2 = 0 \times 10^{11}$ cm^{-2}.

subband. The ψ_{n_1} and ψ_{n_2} wave functions for each conduction subband in the z axis will be given by the nonlinear Schrödinger equations Cruz (2011)

$$i\hbar \frac{\partial}{\partial t} \psi_{n_1}(z,t) = \left[-\frac{\hbar^2}{2m^*} \frac{\partial^2}{\partial z^2} + V(z) + V_H \left(| \psi_{n_1} |^2, | \psi_{n_2} |^2 \right) \right] \psi_{n_1}(z,t), \tag{13}$$

$$i\hbar \frac{\partial}{\partial t} \psi_{n_2}(z,t) = \left[-\frac{\hbar^2}{2m^*} \frac{\partial^2}{\partial z^2} + V(z) + V_H \left(| \psi_{n_1} |^2, | \psi_{n_2} |^2 \right) \right] \psi_{n_2}(z,t), \tag{14}$$

where the subscripts n_1 and n_2 refer to the subband number, respectively, and $V(z)$ is the potential due to the quantum wells. The m^* is the electron effective mass. V_H is the Hartree potential given by the electron-electron interaction in the heterostructure region. Such a many-body potential is given by Poisson's equation Cruz (2002)

$$\frac{\partial^2}{\partial z^2} V_H(z,t) = -\frac{e^2}{\varepsilon} \left[n_1 \left| \psi_{n_1}(z,t) \right|^2 + n_2 \left| \psi_{n_2}(z,t) \right|^2 \right], \tag{15}$$

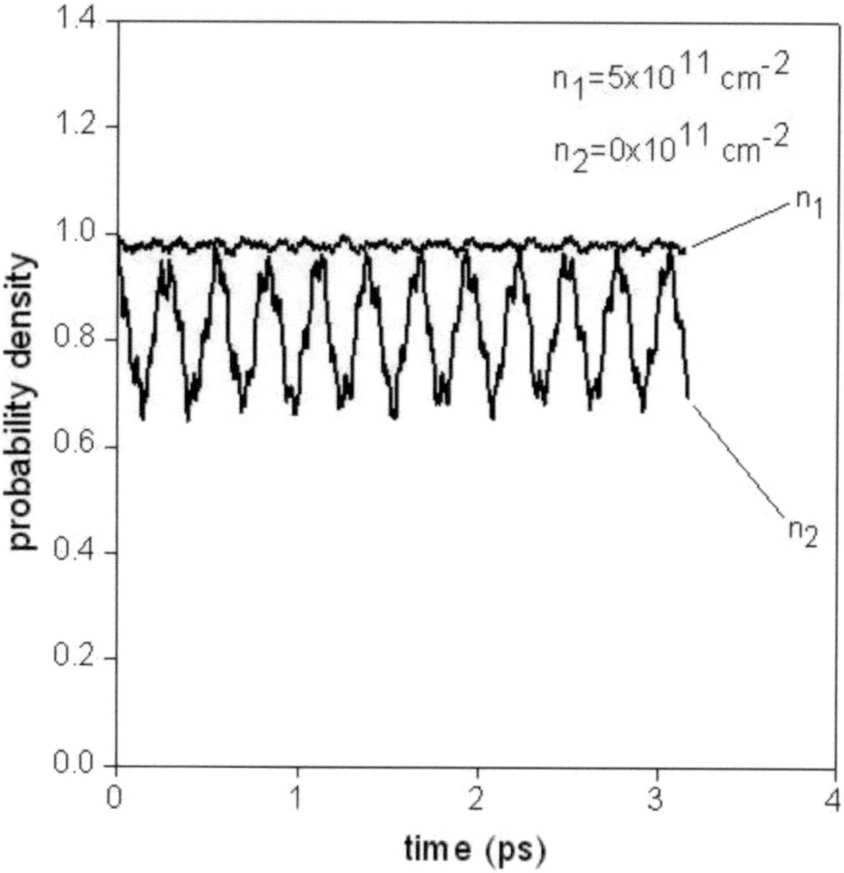

Fig. 7. Probability density in the left quantum well versus time at different carrier densities. $n_1 = 5 \times 10^{11}$ cm^{-2} and $n_2 = 0 \times 10^{11}$ cm^{-2}.

where ε is the GaAs dielectric constant and n_1 and n_2 are the carrier sheet densities in each subband. Considering the Fermi energy ε_F, the carrier densities can be easily calculated. If $\varepsilon_F < \varepsilon_2$, we have

$$n_1 = (\varepsilon_F - \varepsilon_1)\rho_0 \tag{16}$$

and $n_2 = 0$ and if $\varepsilon_F > \varepsilon_2$, we have

$$n_1 = (\varepsilon_F - \varepsilon_1)\rho_0 \tag{17}$$

and

$$n_2 = (\varepsilon_F - \varepsilon_2)\rho_0. \tag{18}$$

In such a case, L is the quantum well width,

$$\varepsilon_n = \hbar^2 n^2 \pi^2 / 2m^* L^2 \tag{19}$$

approaches the quantum well energy levels and

$$\rho_0 = m^* / \pi\hbar^2 \tag{20}$$

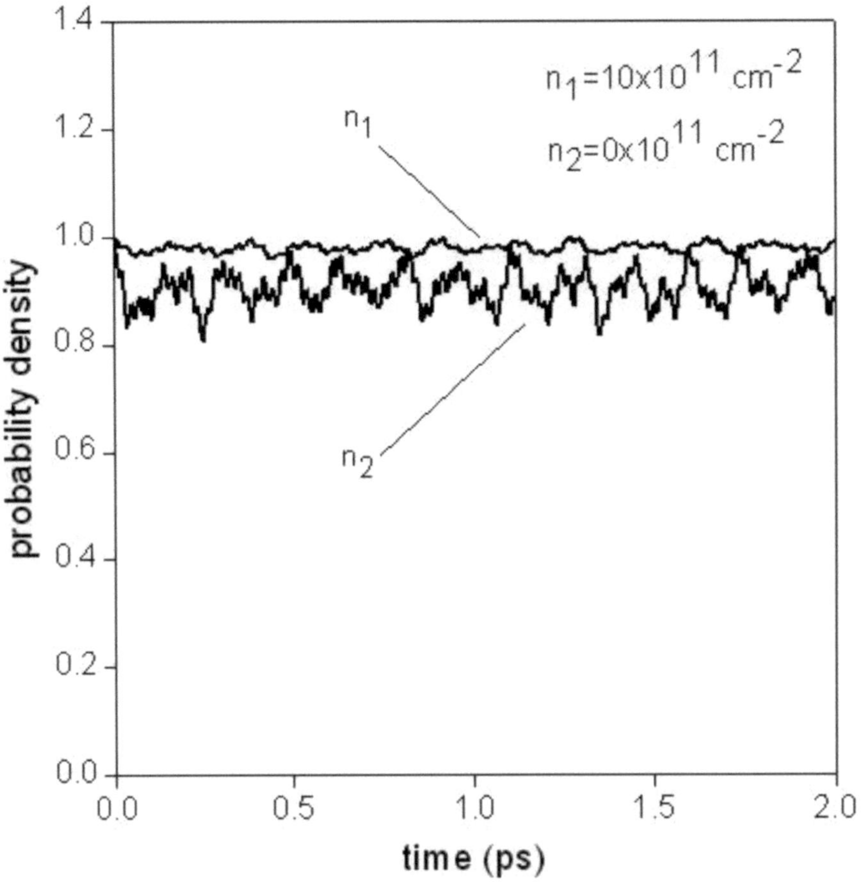

Fig. 8. Probability density in the left quantum well versus time at different carrier densities. $n_1 = 10 \times 10^{11}$ cm^{-2} and $n_2 = 0 \times 10^{11}$ cm^{-2}.

is the two dimensional density of states at zero temperature.

Now we discretize time by a superscript ϑ and spatial position in the subbands by a subscript ξ and φ, respectively. Thus,

$$\psi_{n_1} \to \psi_{\xi}^{\vartheta} \tag{21}$$

and

$$\psi_{n_2} \to \psi_{\varphi}^{\vartheta}. \tag{22}$$

The various z values become $\xi\delta z$ in the conduction band and $\varphi\delta z$, where δz is the mesh width.

Similarly, the time variable takes the values $\vartheta\delta t$, where δt is the time step. We have used a unitary propagation scheme for the evolution operator obtaining a tridiagonal linear system that can be solved by using the split-step method Cruz (2002).

In the split-step approach, both wave packets are advanced in time steps δt short enough that the algorithm

$$e^{-i\delta t T_H/2} e^{-i\delta t U} e^{-i\delta t T_H/2} \tag{23}$$

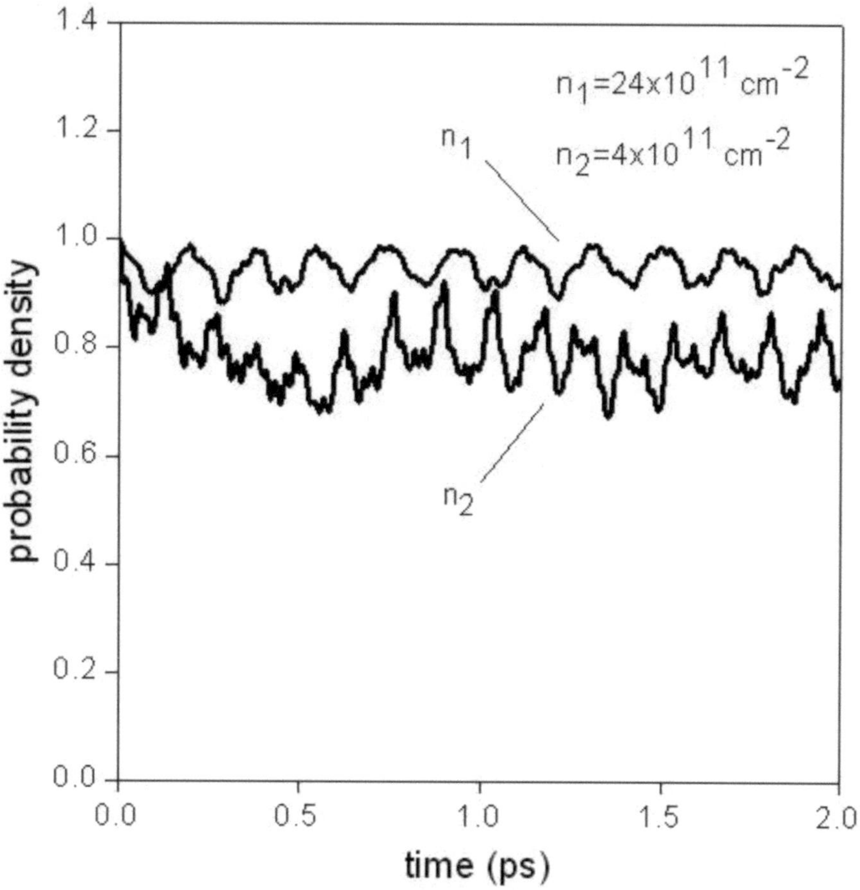

Fig. 9. Probability density in the left quantum well versus time at different carrier densities. $n_1 = 24 \times 10^{11}$ cm^{-2} and $n_2 = 4 \times 10^{11}$ cm^{-2}.

can be applied to the generator. T_H and U are the Hamiltonian kinetic and potential terms.

Then, Poisson's equation associated with V_H is solved using another tridiagonal numerical method for each δt value. In each time step δt, the algorithm propagates the wave packets freely for $\delta t/2$, applies the full potential interaction, then propagates freely for the remaining $\delta t/2$. The split-step algorithm is stable and norm preserving and it is well suited to time-dependent Hamiltonian problems.

We have numerically integrated Eqs. (13), (14) and (15) using $n_1 = 3.0 \times 10^{11}$ cm^{-2} and $n_2 = 0.0 \times 10^{11}$ cm^{-2} carrier densities. In our calculations, we shall consider a GaAs double quantum dot system. We have assumed that both ψ_{n_1} and ψ_{n_2} wave functions are initially created in the center of the left quantum well at $t = 0$ in our model (Fig. 1).

Then, the equations are numerically solved using a spatial mesh size of 0.5Å, a time mesh size of 0.2 a.u and a finite box (5,000Å) large enough as to neglect border effects. The electron effective-mass is taken to be 0.067m$_0$ and L=150 Å. The barrier thickness is 20 Å.

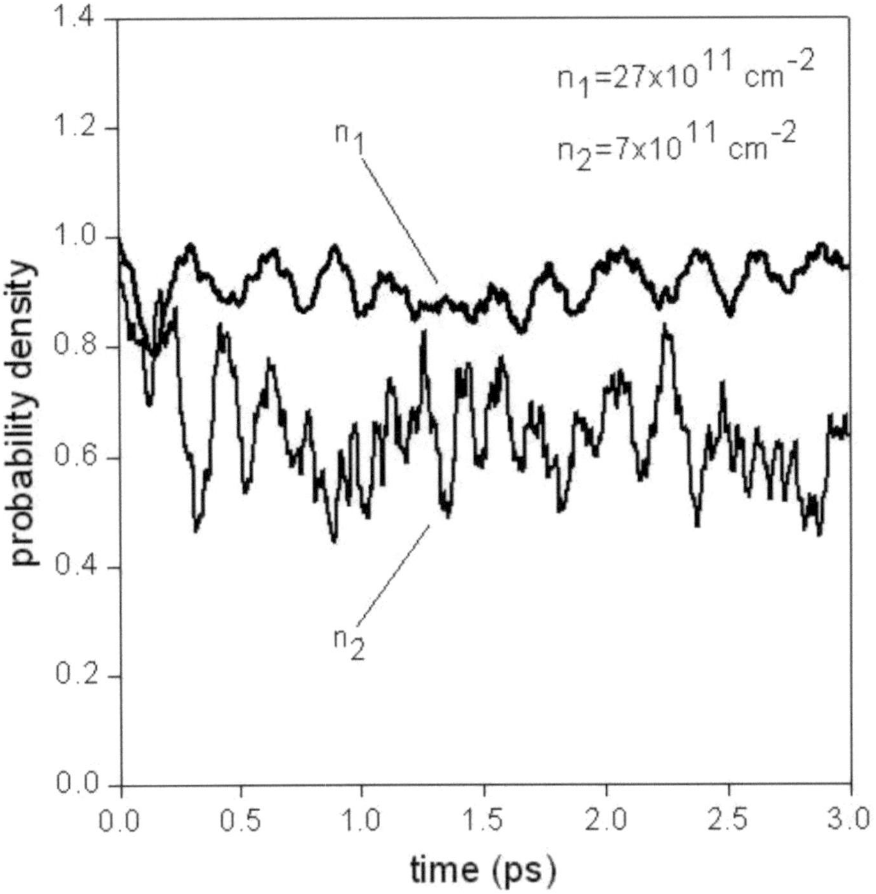

Fig. 10. Probability density in the left quantum well versus time at different carrier densities. $n_1 = 27 \times 10^{11}$ cm^{-2} and $n_2 = 7 \times 10^{11}$ cm^{-2}.

3. Results

The numerical integration in time allows us to obtain the carrier probability, P, in a defined semiconductor region $[a, b]$ and electron subband at any time t

$$P_{n_1,n_2}(t) = \int_a^b dz\, |\psi_{n_1,n_2}(z,t)|^2, \tag{24}$$

where $[a,b]$ are the quantum well limits. In Fig. 4-10 we have plotted the electron probability density in the left quantum well versus time at different electronic sheet densities.

The charge density values were obtained through Eq. (24). The existence of tunneling oscillations between both quantum wells at low densities is shown in Fig. 4. In Fig. 4 it is found that the amplitude of the oscillating charge density is approximately equal to 1 at resonant condition.

The electron energy levels of both wells are exactly aligned at $n_1 = 0.0 \times 10^{11}$cm^{-2} and $n_2 = 0.0 \times 10^{11}$cm^{-2} (Fig. 1) in the conduction band. In our case, the total charge density

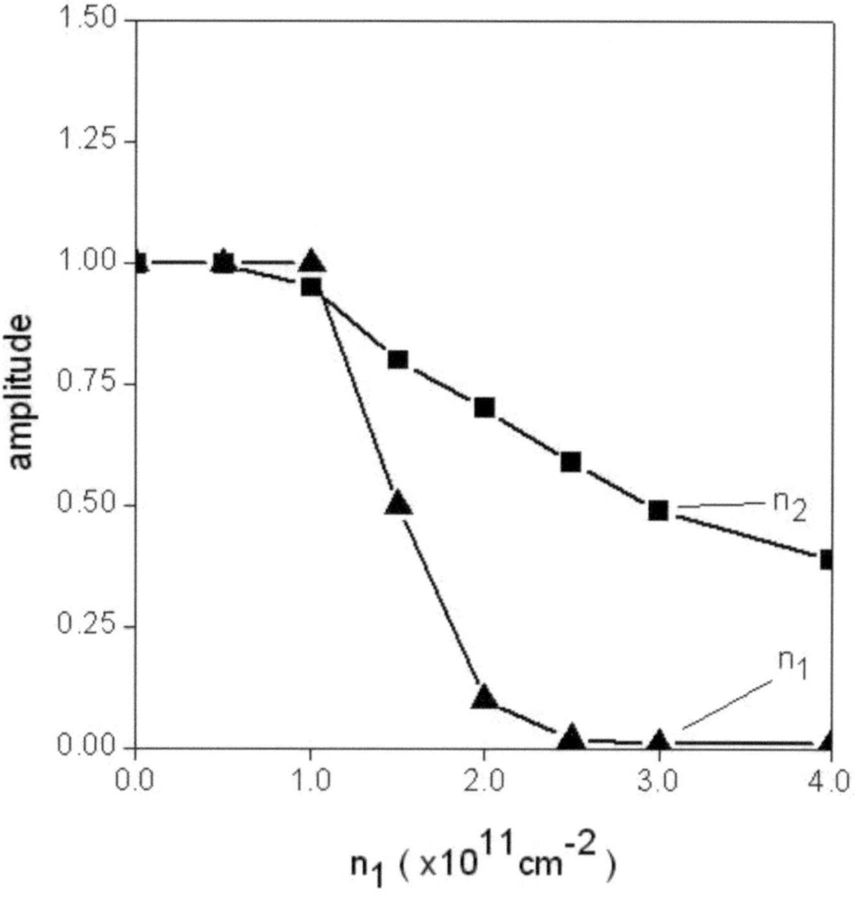

Fig. 11. Amplitude of the tunneling oscillations versus carrier sheet density. Triangles: first subband. Squares: second subband.

will oscillate between both wells with a certain tunneling period due to $n_1 \approx 0$ ($n_1 = 0.1 \times 10^{11} cm^{-2}$).

The level splitting between both quantum wells is proportional to the inverse of the tunneling period. The subsequent evolution of the wave function will basically depend on such a value of the level splitting. However, the quantum well eigenvalues are not aligned for a higher n_1 value, Fig. 5. Then, the amplitude of the oscillating charge is not always equal to 1.

When the n_1 wave function is in the right quantum well, P_{n_2} is never equal to 1, see the arrow (1) in Fig. 5. And when the n_1 wave function is in the left quantum well, P_{n_2} is never equal to 0, see the arrow (2) in Fig. 5.

In such a case, the charge dynamically trapped in the double-well system produces a reaction field which modifies the P_{n_2} value of the charge density oscillations for both wave packets. As a result, the averaged amplitude of the oscillating charge density is never equal to 1, Fig. 5.

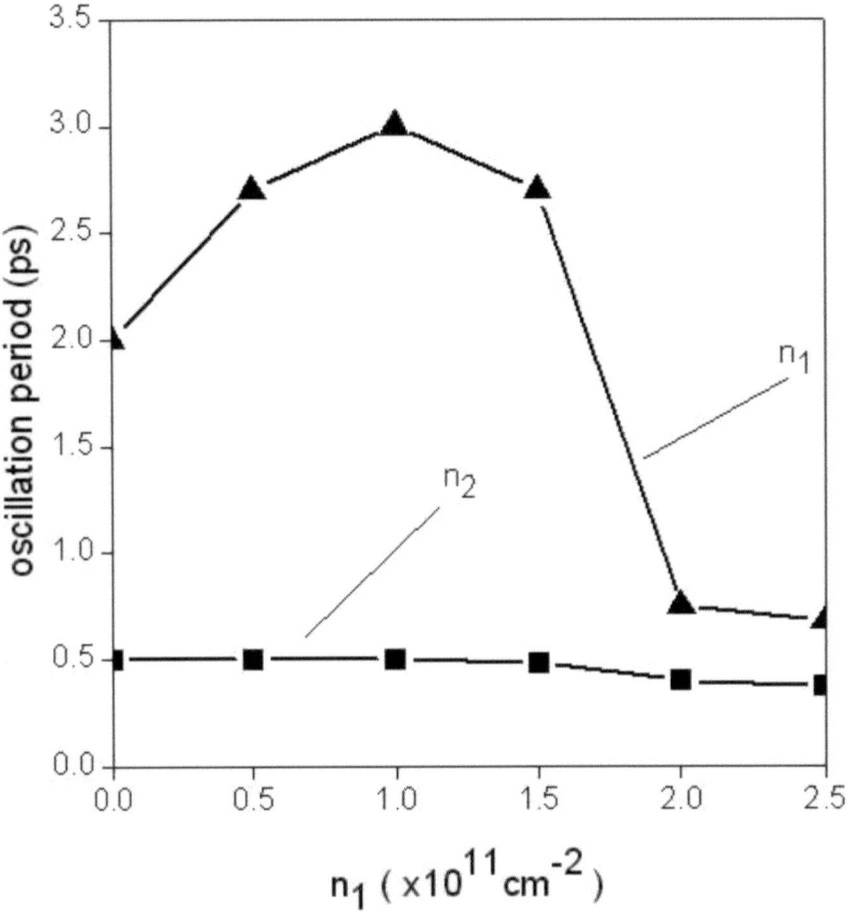

Fig. 12. Period of the tunneling oscillations versus carrier sheet density. Triangles: first subband. Squares: second subband.

Now we plot the averaged amplitude of the tunneling oscillations versus n_1 for low ε_F values, i.e., $\varepsilon_F < \varepsilon_2$ in Fig. 11. At $n_2 = 0.0 \times 10^{11}\text{cm}^{-2}$, it is found that the amplitude of the tunneling oscillations for both wave packets decreases as we increase n_1.

Such a new nonlinear effect is given by the n_1 charge density. The n_2 curve decrease is less than that obtained in the n_1 case in Fig. 11. Such a result can be easily explained as follows. If the potential difference between both wells is higher than the level splitting, the resonant condition is not obtained, and then the tunnelling process is not allowed.

The level splitting in the first subband is much smaller than in the second subband case due to the different barrier transparency, Fig. 1. We can notice that the barrier transparency increases as we increase the energy in a double quantum well. Then, the nonlinear effects are more important in the n_1 case.

We plot the period of the tunneling oscillations versus the n_1 carrier sheet density at $n_2 = 0.0 \times 10^{11}\text{cm}^{-2}$ in Fig. 12. It is found that the oscillation period of the first subband is always

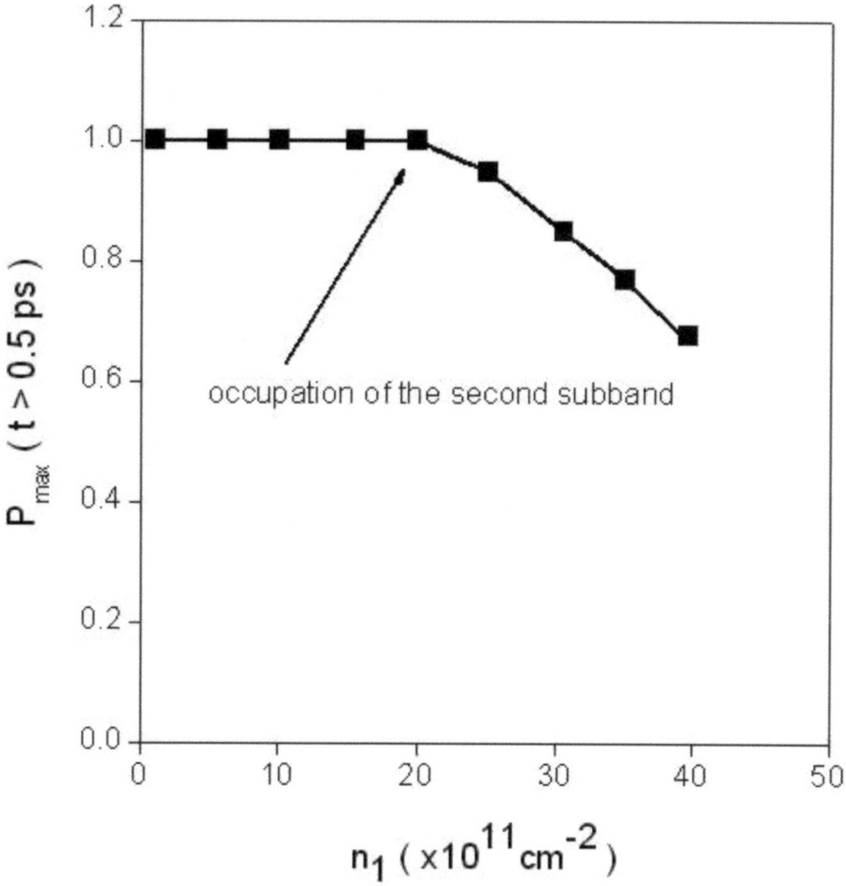

Fig. 13. The maximum probability density in the left quantum well for the second subband after a small initial period ($t > 0.5ps$)

higher than in the n_2 case. Such a result can be explained as follows. We know that the electron tunneling time between two quantum wells decreases as we increase energy.

The electrons in the second subband have higher energy, and an smaller oscillation period, than the n_1 electrons. The tunneling time in the first subband is strongly affected by the n_1 charge density. As a consequence, the nonlinear effects are more important in the n_1 case due to the level splitting in the first subband is much smaller.

In addition to this, and if the number of electrons is large enough, both electron subbands can be occupied. In such a case, we have intersubband interaction, i.e., $n_1 > 0$ and $n_2 > 0$ ($\varepsilon_F > \varepsilon_2$) in Fig. 9-10. Important nonlinear effect in the tunneling oscillations between both quantum wells, which modifies the dynamical evolution of the system, are shown.

The time-dependent evolution of the electron wave packets is strongly modified due to the repulsive intersubband interaction between both wave functions at $\varepsilon_F > \varepsilon_2$ values. We have two different wave functions for two electron groups that interact between each other.

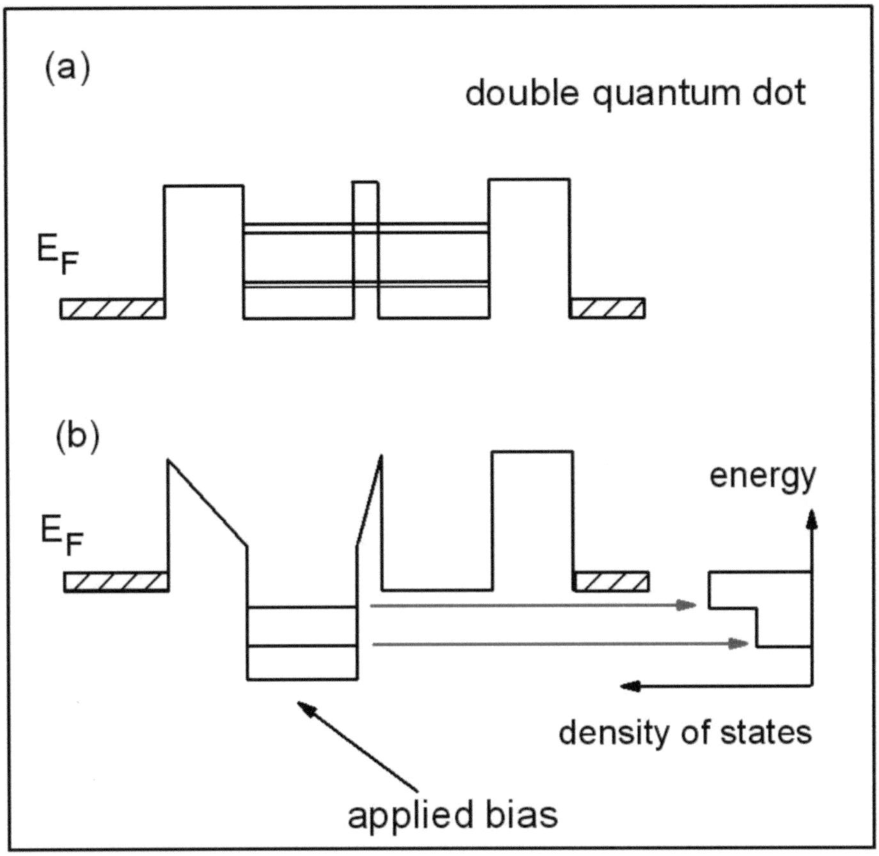

Fig. 14. A schematic illustration of the proposed experiment (a) Double quantum dot system in absence of external bias. (b) The electrons in the left reservoir can tunnel into the left dot when an external voltage is applied to the left quantum well. The density of states is filled up to the Fermi energy ε_F. As a consequence, electrons can be initially injected in the left quantum dot.

The charge dynamically trapped in the double-well system produces a reaction field which modifies the form of the probability curves for both ψ_{n_1} and ψ_{n_2}. As a result we have found important nonlinear effects in the tunneling dynamics for both subbands in Fig. 10.

The amplitude of the oscillating charge density is never equal to 1 in the second subband at high n_1 and n_2 values after a small initial period (t >0.5 ps). We plot P_{max} the maximum probability density in the left quantum well for the second subband after a small initial period ($t > 0.5$ ps) in Fig. 13. It is shown that the P_{max} value is decreased as we increase n_1.

As we increase both n_1 and n_2 values, the nonlinear effects due to the repulsive intersubband interaction are increased. At $\varepsilon_F > \varepsilon_2$ values, it is found that the symmetry of the oscillations is broken due to the nonlinear effects (Fig. 10).

As a result, it is shown the possibility of suppression of the tunneling oscillations in the double quantum well system in the $\varepsilon_F > \varepsilon_2$ regime ($n_1 > 20.0 \times 10^{11}$ cm^{-2}), Fig. 13. We explain this effect by considering our nonlinear effective-mass Schrödinger equations. In absence of intersubband interaction, i.e., $n_1 > 0$ and $n_2 = 0$, we know that the maximum P value of the oscillating charge density is approximately equal to 1 at low n_1 values, Fig. 4.

In such a case, the nonlinear effects are generated by a single charge distribution. However, and at high n_1 and n_2 values, we have a reaction field generated by two charge distributions. If two subbands are occupied, important nonlinear effects in the carrier dynamics are obtained (Fig. 10).

As it is shown in Fig. 1, electrons can be initially distributed in both subbands. The electrons in the left reservoir can tunnel into the left dot, Fig. 14, when an external voltage is applied to the left quantum well. Then, the quantum states in the left quantum well are filled up to the Fermi energy. If we now switch off the applied voltage, Fig. 14, we obtain electrons distributed in both subbands that are localized in the left quantum well.

The initial wave functions $\psi_{n_1}(t = 0)$ and $\psi_{n_2}(t = 0)$ correspond to quantum well eigenstates in the left dot. In such experiment, the superposition of both symmetric and antisymmetric quantum-well eigenstates in the conduction band leads to coherent tunneling between both quantum wells. We have two different charge densities that oscillate with different tunneling periods.

4. Conclusion

In this work, we have studied the post-measurement dynamics in a double quantum dot system considering two subband wave packets. We have numerically integrated in space and time the effective-mass Schrödringer equation for two electron gases in a double quantum dot system.

We found two time-varying moments in the nanostructure with two different frequencies. In addition, it is found important nonlinear effects if two electron subbands are occupied. The symmetry of the tunneling oscillation can be broken due to nonlinear effects at high charge density values.

5. References

Astley, M. R.; Kataoka, M.; Ford, C. B. J.; Barnes, C. H. W., Anderson, D., Jones, G. A. C.; Farrer, I.; Ritchie, D. A. & Pepper, M. (2007). Energy Dependent Tunneling from Few-Electron Dynamic Quantum Dots. *Physical Review Letters*. Vol. 99, 156802-156806, ISBN 0031-9007/07/99(15)/156802(4)

Cruz, H. (2002). Tunneling and time-dependent magnetic phase transitions in a bilayer electron system. *Physical Review B*, Vol. 65, 245313-245318, ISBN 0163-1829/2002/65(24)/245313(5)

Cruz, H. & Luis, D. (2011). Coulomb effects and sub-band tunneling in quantum wells. *Journal of Applied Physics*, Vol. 109, 073725-073729, ISBN 0021-8979/2011/109(7)/073725(5)

Ferreira, G. P.; Freire, H. J. P. & Egues, J. C. (2010). Many-Body Effects on the ρ_{xx} Ringlike Structures in Two-Subband Wells. *Physical Review Letters*. Vol. 104, 066803-066807, ISBN 0031-9007/10/104(6)/066803(4)

Shabami, J.; Lin, Y. & Shayegan, M. (2010). Quantum Coherence in a One-Electron Semiconductor Charge Qubit. *Physical Review Letters*, Vol. 105, 246804-246808, ISBN 0031-9007/10/105(24)/246804(4)

Van der Wiel, W. G.; De Franceschi, S.; Elzerman J. M.; Fujisawa, T.; Tarucha, S. & Konwenhoven, L. P. (2003). Electron transport through double quantum dots. *Review of Modern Physics*, Vol. 75, No. 1, 1-22, ISBN 0034-6861/2003/75(1)/1(22)

3

Block Diagram Programming of Quantum Dot Sources and Infrared Photodetectors for Gamma Radiation Detection Through VisSim

Mohamed S. El-Tokhy[1], Imbaby I. Mahmoud[1], and Hussein A. Konber[2]
[1]*Engineering Department, NRC, Atomic Energy Authority, Inshas, Cairo*
[2]*Electrical Engineering Department, Al Azhar University, Nasr City, Cairo*
Egypt

1. Introduction

Quantum dots (QDs) have attracted tremendous interest over the last few years for a large variety of applications ranging from optoelectronic through photocatalytic to biomedical, including applications as nanophosphors in light emitting diodes LEDs [1]. QDs have been suggested as scintillators for detection of alpha particles and gamma-rays [1-4]. Quantum dots offer an improvement on scintillator technology in that the size of the phosphorescent material would not be restrained by crystal growth. The dots would be suspended in a transparent matrix that could be as large as desired. Moreover, the output of the QDs is a function of their dimensions. Therefore, they could be produced to emit light at wavelengths suitable for detection by avalanche photodiodes. That offer higher quantum efficiencies than photomultiplier tubes. Therefore, the sensitivity of the scintillator detector is increased [5]. To our knowledge, still models describing QD sources as gamma detection are not devised yet. In this chapter, we report on the effects of gamma irradiation on the different properties of CdSe/ZnS QDs. These characteristics are optical gain, power, population inversion and photon density. Furthermore, a new semiconductor scintillator detector was studied in which high-energy gamma radiation produces electron-hole pairs in a direct-gap semiconductor material that subsequently undergo interband recombination, producing infrared light to be registered by a photo-detector [6]. Therefore, quantum dot infrared photodetectors (QDIPs) can be used for this purpose. This chapter presents a method to evaluate, study, and improve the performance of quantum dot sources and infrared photodetectors as gamma radiation detection. This chapter is organized as follows: Section 3.2 presents the basics of VisSim simulator. QD devices as a detector are illustrated in Section 3.3. The proposed models of both quantum dot devices are represented in Section 3.4. Discussion and results are summarized in Section 3.5.

1.2 VisSim simulator

VisSim simulation is one of the most widely used environments in operations research and management science, and by all indications its popularity is on the increase. The goal of using VisSim modeling and analysis is to give an up-to-date treatment of all the important aspects of a simulation study, including modeling, simulation languages, validation, and output data analysis. However, most real world-systems are too complex to allow realistic models to be evaluated analytically, and these models must be studied by means of VisSim simulation. In VisSim we use a computer to evaluate a model numerically over a time period of interest, and data are gathered to estimate the desired true characteristics of the model. In addition we have tried to represent the operation of the QD sources and photodetectors in a manner understandable to a person having only a basic familiarity with its main behaviors such as optical gain, power, output photon densities, dark current, photocurrent, and detectivity. It can be a powerful supplement to traditional design techniques.

System engineering program such as Mathsoft's VisSim employ the Graphical User Interface (GUI) concept. In order to, display systems of differential equations as feedback systems. Additionally, if the relationships which compose the model are simple enough, it may be possible to use VisSim simulation model. Then, assume numerical integration using the methods of Laplace transforms for differential equations, hence the appearance of the symbols 1/S, to indicate integration. Though these packages integration is used because it is an inherently more stable numerical operation. The first step in converting from equation to block diagram form is thus to integrate both sides of differential equation. The operations of addition, multiplication, integration and so on are performed by blocks operators, as shown in the following models. Besides, reading the block diagram is from left to right.

The following are some possible reasons for the widespread popularity of the VisSim simulation [7-10]:

1. Most complex, real-world systems with stochastic elements cannot be accurately described by a mathematical model which can be evaluated analytically. Thus, VisSim simulation is often the main type of investigation possible.
2. VisSim Simulation allows one to estimate the performance of an exciting system under some projected set of operation conditions.
3. Alternative proposed system designs (or alternative operating policies for a single system) can be compared via VisSim simulation to see which best meets specified requirements.
4. In VisSim simulation we can maintain much better control over experimental conditions than would generally be possible when experimenting with the system itself.
5. VisSim Simulation allows us to study a system with along time frame, or alternatively to study the detailed workings of a system in expanded time.

Our goal in this chapter is to evaluate the performance of QDs devices by using VisSim environment along with the block diagram programming procedures.

1.3 QD devices as a detector

Quantum dots have been used in a wide variety of applications. A key advantage of these particles is that their optical properties depend predictably on size, which enables tuning of the emission wavelength [11]. QD devices as a source and infrared photodetectors can be used to detect gamma radiation [2, 4, 6]. Moreover, the ability of semiconductor quantum dots to convert alpha radiation into visible photons was demonstrated in [2].

In this chapter, we report on the scintillation of quantum dots sources and infrared photodetectors under gamma-ray irradiation.

Semiconductor scintillation gamma radiation detector based on QD will be discussed in which the gamma-ray absorbing semiconductor body is impregnated with multiple small direct gap semiconductor inclusions of band gap slightly narrower than that of the body. If the typical distance between them is smaller than the diffusion length of carriers in the body material, the photo-generated electrons and holes will recombine inside the impregnations and produce scintillating radiation to which the wide gap body is essentially transparent [6]. Furthermore, the quantum dot lasers (QDLs) under gamma radiation is characterized by changes in threshold current, external slope efficiency and light output [12].

For infrared (IR) detection, most of QDIPs are based on vertical heterostructures consisting of two dimensional arrays of QDs separated by the barrier layers. The QD structures serving as the photodetector active region, where IR radiation is absorbed, are sandwiched between heavily doped emitter and collector contact layers. The active region can be either doped (with dopants of the same type as the contact layers) or undoped. Schematic view of vertical QDIPs device structures is in [13]. The absorption of IR is associated with the electron intersubband transitions from bound states in QDs into continuum states above the barriers or into excited quasi-bound states near the barrier top. The bound-to-continuum transitions or bound-to-quasi-bound transitions followed by fast escape into the continuum result in the photoionization of QDs and the appearance of mobile electrons. Bound electrons accumulated in QDs can create a significant space charge in the active region. In this chapter, we extended the same physical characteristics to study QD devices under gamma radiation.

1.4 The proposed models of quantum dot devices

1.4.1 The proposed simulator of quantum dot laser under gamma radiation

Improved radiation detection using quantum dot semiconductor composites is demonstrated [4, 14]. The quantum confinement effects enabled the QDs to be used with a wide variety of detectors. The inorganic nanocrystalline quantum dots will be demonstrated as a powerful method in preparing scintillating devices. In gamma-ray detection, our first goal is to employ nanomaterials in the form of QD based mixed matrices to achieve scintillation output several times over that from NaI(Tl) crystals. Moreover, block diagram models are used to represent carrier densities in the Subbands, population inversion process, optical power, and gain. Our goal from this study is to obtain a device that has high performance with optimal operating conditions.

1.4.1.1 Proposed simulator of optical wavelength

This subsection introduces a useful block diagram model to predict effect of incident gamma radiation on quantum dot emission wavelength.

The QDs semiconductor composite is designed so that ionizing radiation produces excitations predominantly in the semiconductor QDs. These excitations are subsequently Förster-transferred to organic material. For scintillators, the composite material must be transparent to the emitted photon, and the large Stokes shift of the organic material is essential. For charge-collection devices the composite material must be trap free to allow efficient charge collection. Trap-free, conjugated organic materials are now available [4]. Pure QDs solids are impractical radiation-detection materials because they are not transparent to their emission wavelength and have significant charge-carrier trapping.

If the peak photoluminescence (PL) photon emission energy, E_{PL}, is always lower than the 1s–1s absorbance peak energy, E_{abs}, by an energy shift ($\Delta E_{AP} \sim -0.14$ eV + 0.074 x E_{abs} or ΔE_{AP} ~0.025 eV for CdSe quantum dots when 2 eV < E_{abs} < 2.5 eV), then the wavelength of the PL peak, λ_{PL}, can be calculated as follows [15]

$$\lambda_{PL} = \frac{hc}{E_g + X/R - \Delta E_{AP}} \tag{3.1}$$

where h, c, E_g, X denote plank's constant, speed of light, energy gap, and $X = 0.82\text{x}10^{-7}$ eV-cm for CdSe, respectively. If all Cd precursors have been consumed, the average nanocrystal radius reaches the completion radius, R_c, and PL emission occurs at a completion wavelength, λ_c.

$$\lambda_c = \lambda_{PL}\big|_{R=R_c} \tag{3.2}$$

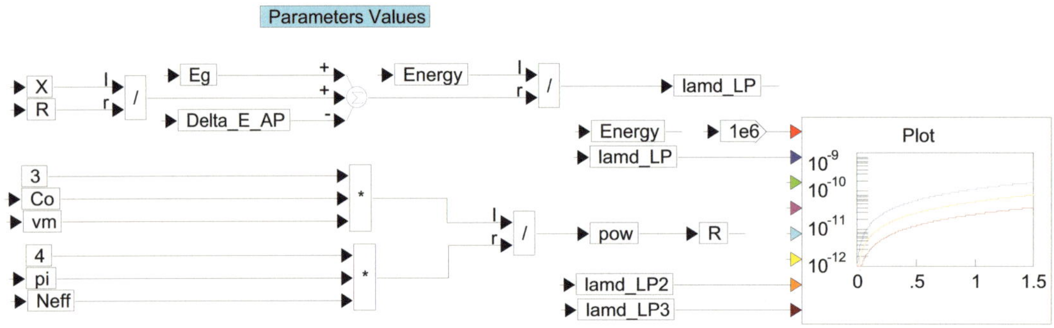

Fig. 3.1. Block diagram model that describes the effect of incident gamma energy on emission wavelength.

Growth of the average quantum dot radius can be written as [15]

$$R_c = \left(\frac{3C_0V_m}{4\pi N_{eff}}\right)^{1/3} \tag{3.3}$$

where C_0, N_{eff}, and V_m denote the original molar concentration of Cd in the batch (in moles cm^{-3}), the effective number of spherical nanocrystals that would contain the number of moles of Cd that have been consumed, and the molar volume, respectively. Block diagram model that describe the effect of γ-energy on the emission wavelength is depicted in Fig. 3.1.

1.4.2 The models of quantum dot infrared photodetectors

A new scintillation-type semiconductor detector was studied in which high-energy radiation produces electron-hole pairs in a direct-gap semiconductor material that subsequently undergo interband recombination, producing infrared light to be registered by a QDIP. Scintillators are not normally made of semiconductor material. The key issue in implementing a semiconductor scintillator is how to make the material essentially transparent to its own infrared light. Consequently, that photons generated deep inside the semiconductor slab could reach its surface without tangible attenuation.

In this subsection, different treatments are applied to model the characteristics of QDIPs than that in [16-17] under gamma radiation. Furthermore, we will utilize a new block diagram model to consider the characteristics of a QDIP under dark and illumination condition.

1.4.2.1 Dark current density block diagram of QDIP

If the numbers of electrons in QDs are sufficiently large, we may assume that these numbers are approximately the same for all QDs in a particular QD array, $N_k^{i,j} = \langle N_k \rangle$ where *i* and *j* are the in-plane indices of QDs, $<N_k>$ denotes average extra carrier number in the QDs and *k* is the index of the QD array. In this case, the distribution of the electric potential $\varphi=\varphi(x,y,z)$ in the active region is governed by the Poisson equation [16]

$$\left(\frac{\partial^2}{\partial x^2}+\frac{\partial^2}{\partial y^2}+\frac{\partial^2}{\partial z^2}\right)\varphi=\frac{4\pi q}{æ}\left(\sum_{i,j,k}\langle N\rangle\delta_{11}(x-x_i)\delta_{11}(y-y_j)\delta_{\perp}(z-z_k)-\rho_D\right) \tag{3.4}$$

where q is the electron charge, æ is the dielectric constant of the material from which the QD is fabricated, $\delta_{11}(x)$, $\delta_{11}(y)$, and $\delta_{\perp}(z)$ are the QD form-factors in lateral (in the QD array plane) and transverse (growth) directions, respectively, x_i and y_j are the in-plane QD coordinates, z_k = kL is the coordinate of the kth QD array (where k = 1, 2, 3, . . . , M and M is the number of the QD arrays in the QDIP), L is the transverse spacing between QDs and ρ_D is the donor concentration in the active region. The form-factors correspond to the lateral and transverse sizes of QDs equal to a_{QD} and l_{QD}, respectively.

Moreover, block diagram model that describes the relation between dark current and structural parameters is implemented through VisSim as depicted in Fig. 3.2.

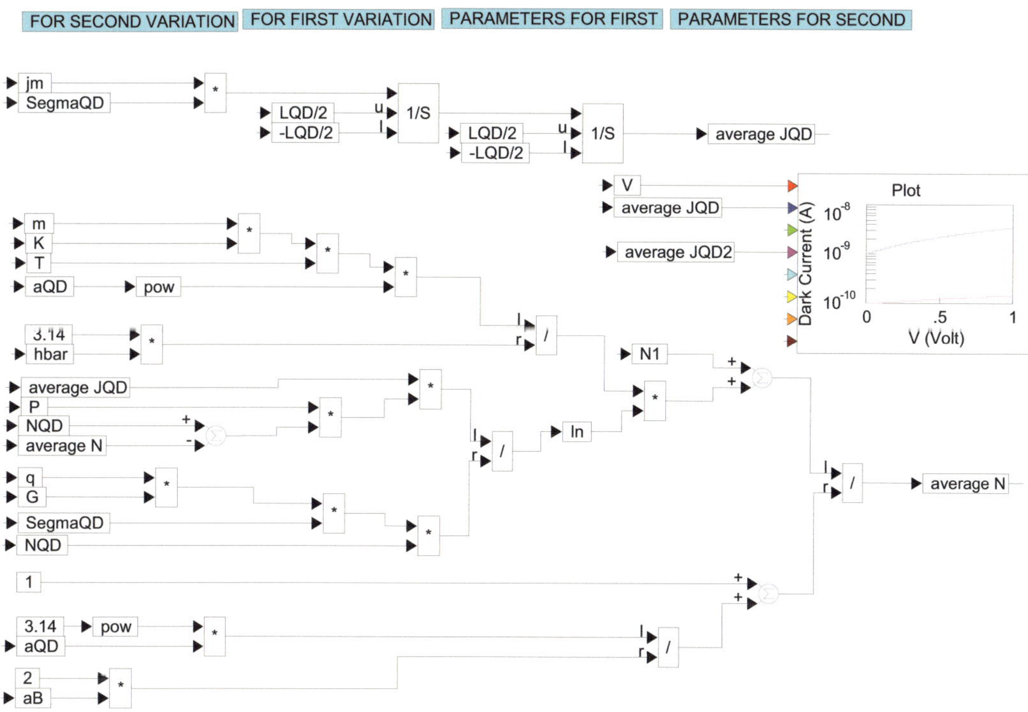

Fig. 3.2. Dark current density block diagram model of QDIPs.

1.4.2.2 Photocurrent density block diagram model of QDIP

We calculate the photocurrent density in QDIPs using a developed device model of [16]. This model takes into account the space charge and the self-consistent electric potential in the QDIP active region, the activation character of the electron capture and its limitation by the Pauli principle, the thermionic electron emission from QDs and thermionic injection of electrons from the emitter contact into the QDIP active region, and the existence of the punctures between QDs. The developed model yields the photocurrent density in a QDIP as a function of its structural parameters. The photocurrent density of QDIP is given by [16-17]

$$J_{Photo} = J_m e^{\frac{q\varphi}{K_B T}} \tag{3.5}$$

where J_m is the maximum current density which can be supplied by the emitter contact. The average photocurrent density of QDIP is given by [16-17]

$$\langle J_{Photo} \rangle = \Sigma_{QD} \int_{-\frac{L_{QD}}{2}}^{\frac{L_{QD}}{2}} \int_{-\frac{L_{QD}}{2}}^{\frac{L_{QD}}{2}} J_{Photo} dxdy = J_m \Sigma_{QD} \int_{-\frac{L_{QD}}{2}}^{\frac{L_{QD}}{2}} \int_{-\frac{L_{QD}}{2}}^{\frac{L_{QD}}{2}} e^{\frac{q\varphi}{K_B T}} dxdy \tag{3.6}$$

Moreover, block diagram model that describes the relation between photocurrent and structural parameters is implemented through VisSim as shown in Fig. 3.3.

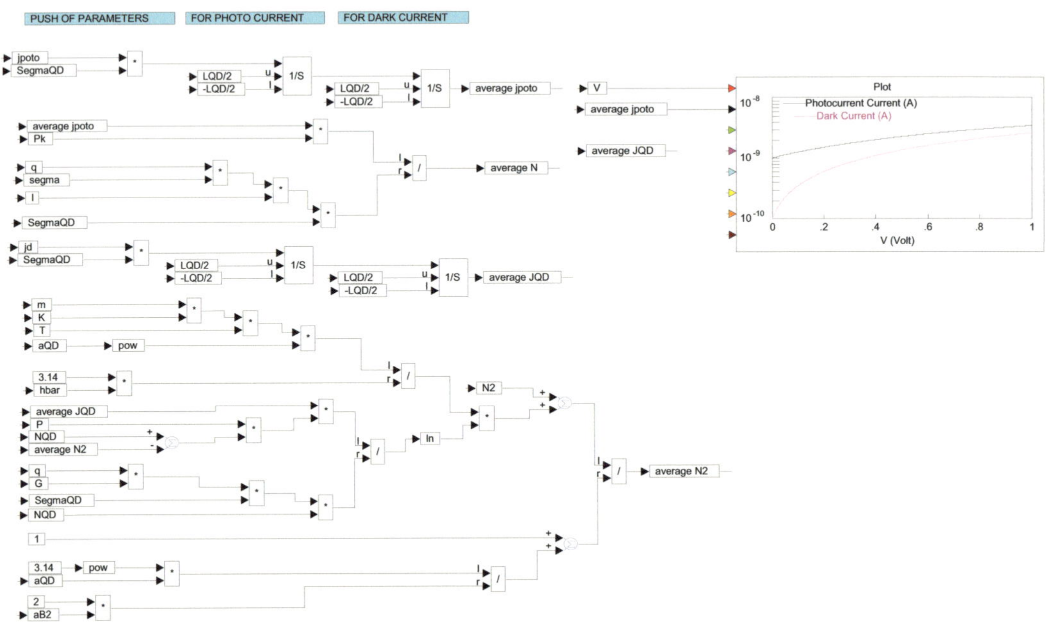

Fig. 3.3. Photocurrent density block diagram model of QDIPs.

1.4.2.3 Detectivity block diagram model of QDIP

The specific detectivity, D, which is a measure of the signal-to-noise ratio of the device, used to characterize QDIPs [18]. It was calculated from the noise density spectra and the peak responsivity [19]. The detectivity of QDIP is determined by the following equation [20-23];

$$D = \frac{R\sqrt{A}}{\sqrt{4\, q\, J_{dark}\, g}} \tag{3.7}$$

Moreover, block diagram model that describes the relation between detectivity and structural parameters is implemented through VisSim as illustrated in Fig. 3.4.

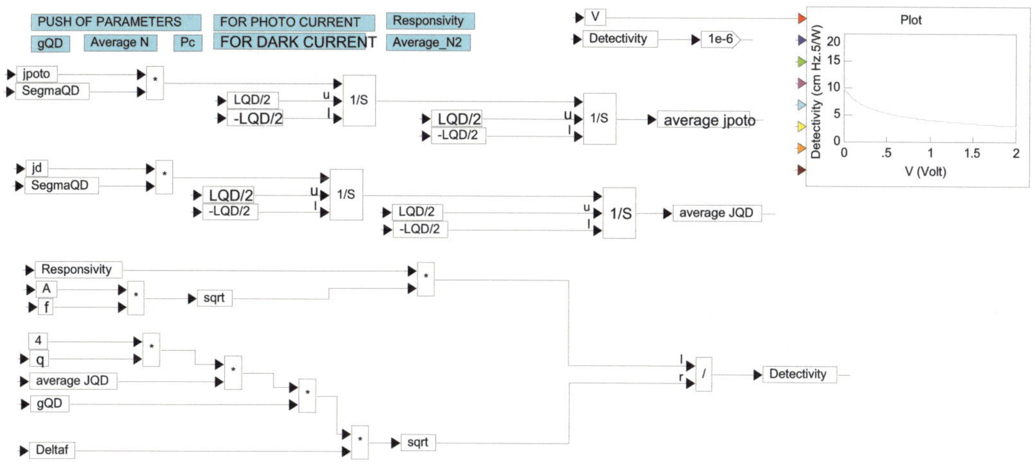

Fig. 3.4 Detectivity block diagram model of QDIPs.

1.5 Results and discussion

In this chapter we are interested with the calculation of different characteristics of quantum dot sources and detectors that can be used as a scintillator for the detection of gamma radiation. The values of the calculations are taken from various references as depicted in Tables 3.1-3.2 [24-31].

n_r=3.5-5	τ_0=10ps	L_{cav}=900µs	E_{cv}=1eV	τ_{wr}=3ns	τ_S=15ps
R_1, R_2=30%,-90%	τ_r=2.8ns	A=6cm^{-1}	τ_p=8.8ps	R=8ns	E_g=0.8eV

Table 3.1 QD source parameters.

Σ_{QD}=(0.1-10)x10^{10} cm^{-2}	N_{QD}=8-10	T=(77-300) K	æ =12	L_{QD}=(40-100) nm
J_m=1.6x10^6 A/cm^2	M=10-70	ρ_D=10^{10} cm^{-3}	M=10-70	a_{QD}=(10-15) nm

Table 3.2 QDIP parameters.

1.5.1 Results of QD sources

The main target of this subsection is to study the various parameters effects on quantum dot devices for gamma radiation detection. Block diagram model using VisSim environment is used for this purpose. Consequently, the current study assists on the fact that QD devices can be used for gamma radiation detection. The effect of incident gamma radiation on the emission wavelength is deeply studied. Moreover, the effect of QD parameters on the optical gain is investigated. The modal gain is determined by calculating the amplified spontaneous emission (ASE) power. We concluded that one of the main advantages of quantum dot (QD) is their wide wavelength range. Therefore, QD sources can be used as an efficient device for gamma detection. The emission wavelength against incident gamma energy at different N_{eff} and V_m is depicted in Figs. 3.5, respectively. The wide range of emission wavelength is one of the main advantages of QD devices. However, the emission wavelength increases with the molar volume as shown in Fig. 3.5. From this study we are concluded that the emission wavelength of QD is wide. Therefore, QD can be used as an efficient device for gamma radiation detection. Furthermore, the intraband free-carrier absorption coefficient in doped semiconductors is roughly proportional to λ^2, which translates into larger optical waveguide losses at the longer wavelengths.

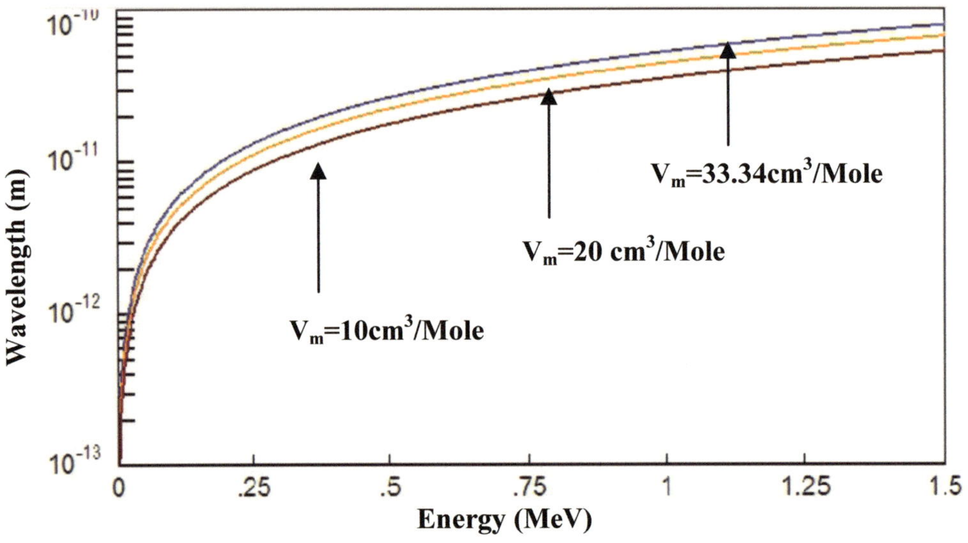

Fig. 3.5. Emission wavelength against incident gamma energy at different values of molar volume

1.5.2 Results of QDIPs

The change of the dark current density with the bias voltage at different sheet densities of QDs is depicted in Fig. 3.6. From this figure, the dark current increases as bias voltage increasing. The main reason for this effect is the non-optimized doping level. Moreover, high agreement between our obtained results with the published results [32] is obtained.

With an increase of Σ_{QD}, the dark current decreases as a result of decreasing number of electrons in quantum dots. The decrease in repulsive potential of charge carriers in quantum dots causes increase in the electron capture probability and decrease in the current gain. In the range of high QD density, Σ_{QD}, the dark current saturates on different levels in dependence on a bias voltage, since the number of electrons in QDs effectively decreases.

Dark current and biasing voltage results at different lateral characteristic size of quantum dot are shown in Fig 3.7. As noted from this figure, the dark current is increased with decreasing the lateral characteristic size.

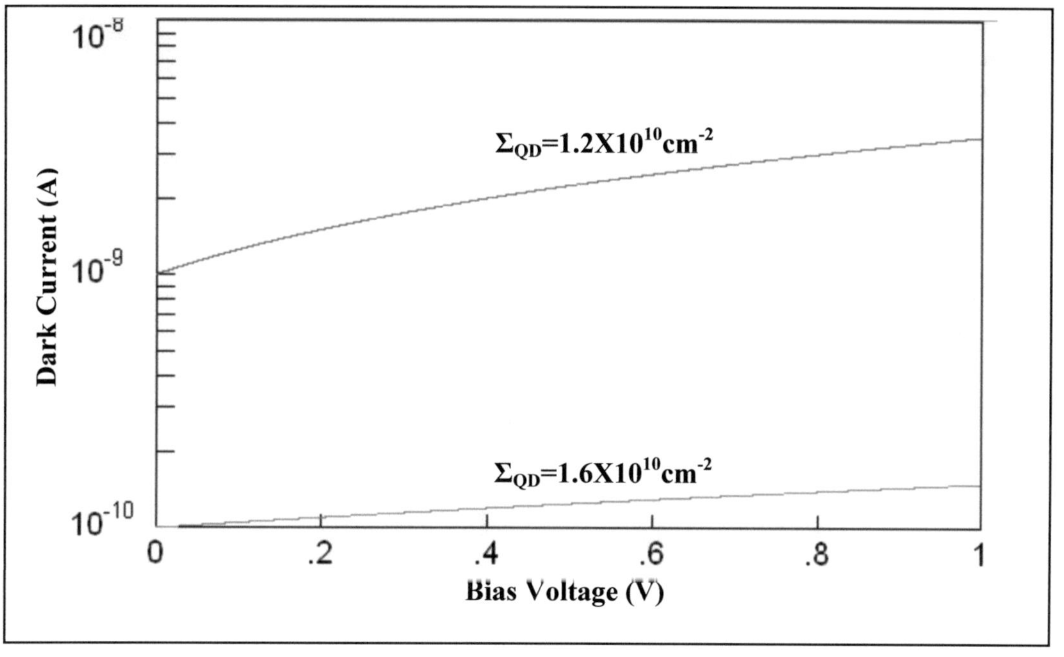

Fig. 3.6. Dark current and biasing voltage results at different quantum dot density of states.

Figures 3.8 depict the change of both dark current and photocurrent with the bias voltage. As seen in this figure, at the beginning the photocurrent is larger than the dark current because there is no thermionic emission of electron and the tunneling process of electron is so small. Then the dark current is sharply increased with the photocurrent, since more electrons are thermally emitted and losses of electrons are increased.

Figures 3.9-3.12, depicts the change of detectivity of QDIP against different parameters: temperature, Σ_{QD}, Σ_{D}, and a_{QD}. As expected from Fig. 3.9, a rapid decrease in detectivity occurs with the increase in a temperature due to contribution of thermal generation and increasing the dark current. From Fig. 3.10, it is obvious that increasing quantum dot density results in increasing detectivity. With increasing Σ_{QD}, the dark current is reduced and the unwanted noise becomes negligible. Hence detectivity is increased. As expected from Figs. 3.11-3.12, a rapid decrease of the detectivity is observed with increasing both Σ_D and the lateral sizes of QDs.

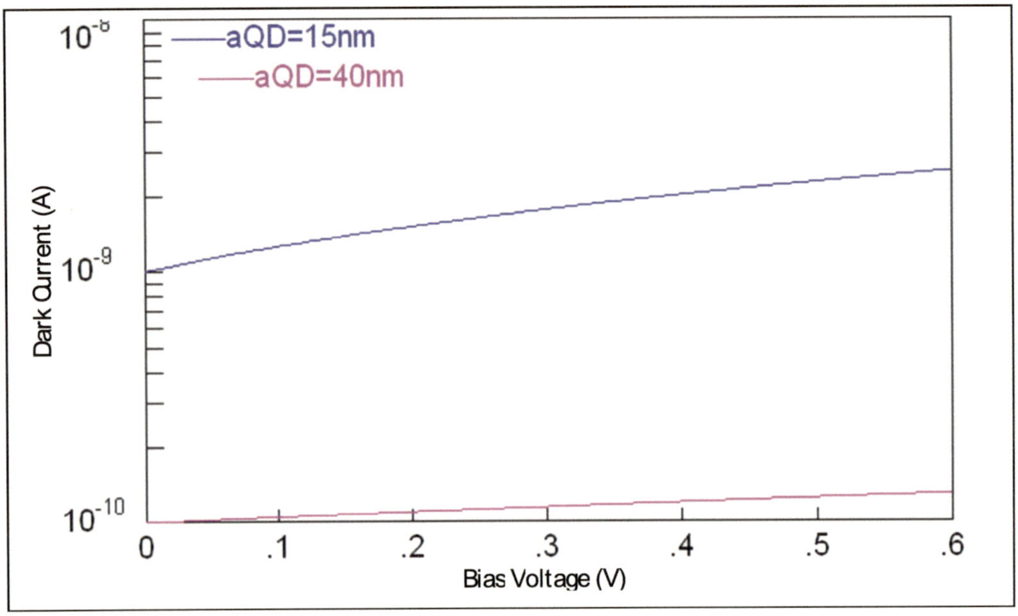

Fig. 3.7. Dark current and biasing voltage results at different lateral characteristic size of quantum dot.

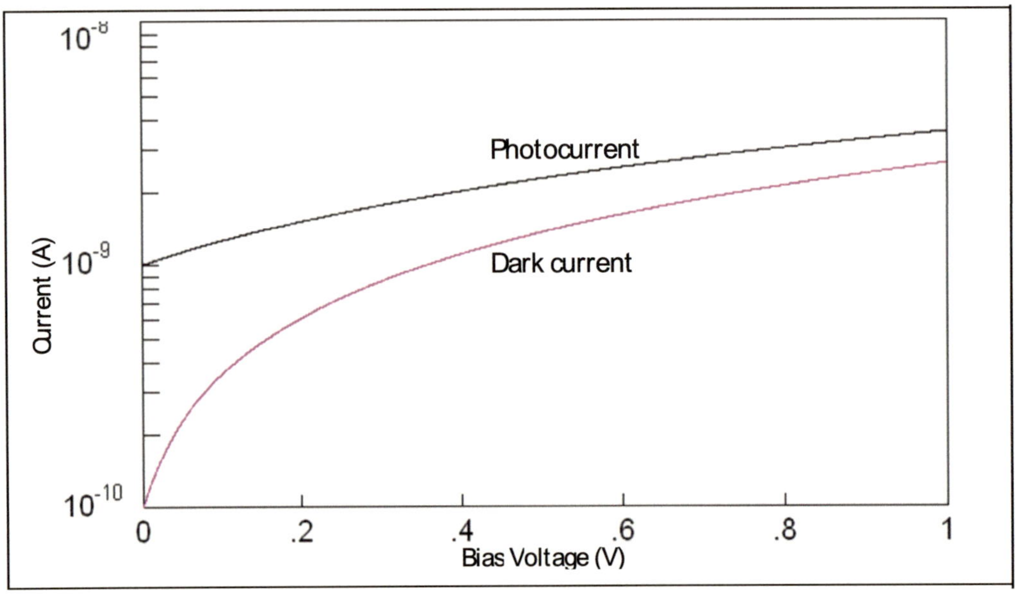

Fig. 3.8. Dark current and photocurrent as a function of the biasing voltage at different quantum dot density of states.

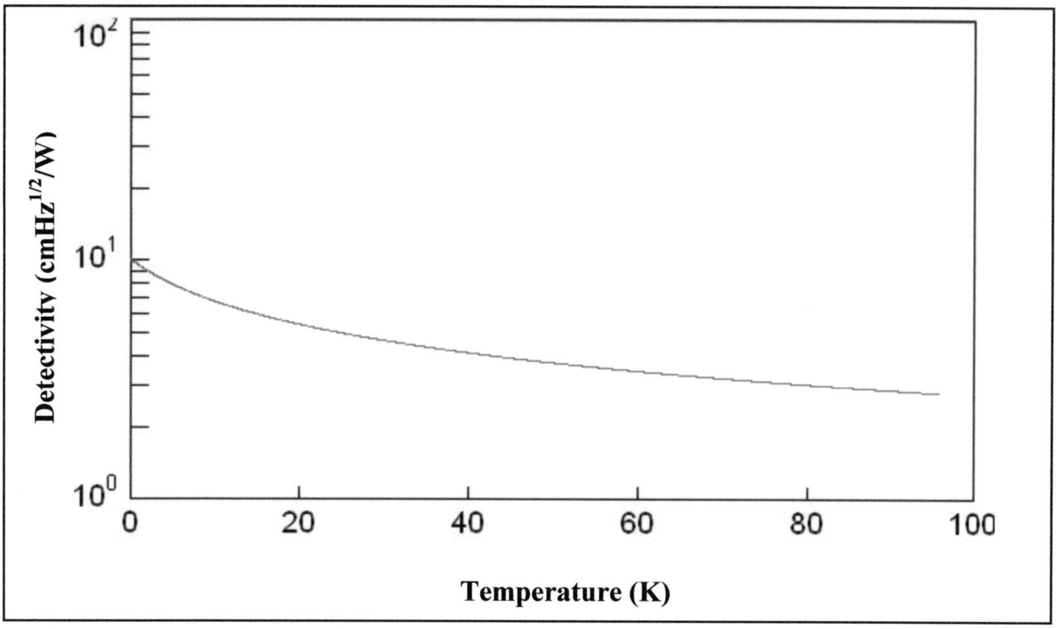

Fig. 3.9. Detectivity against temperature.

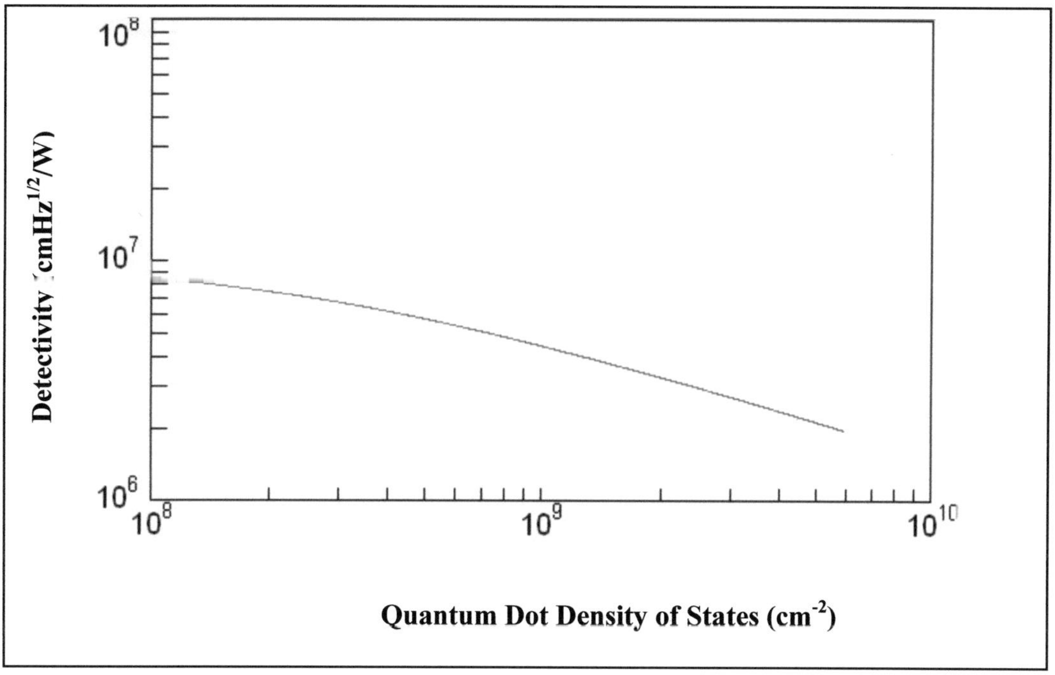

Fig. 3.10. Detectivity against quantum dot density of states.

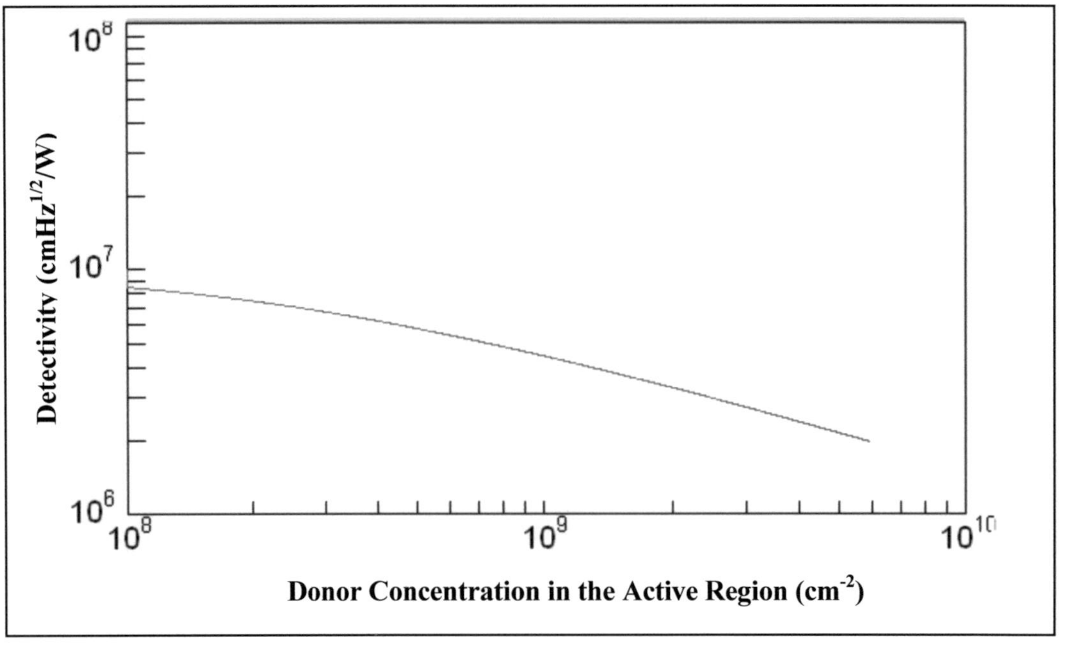

Fig. 3.11. Detectivity against Σ_D.

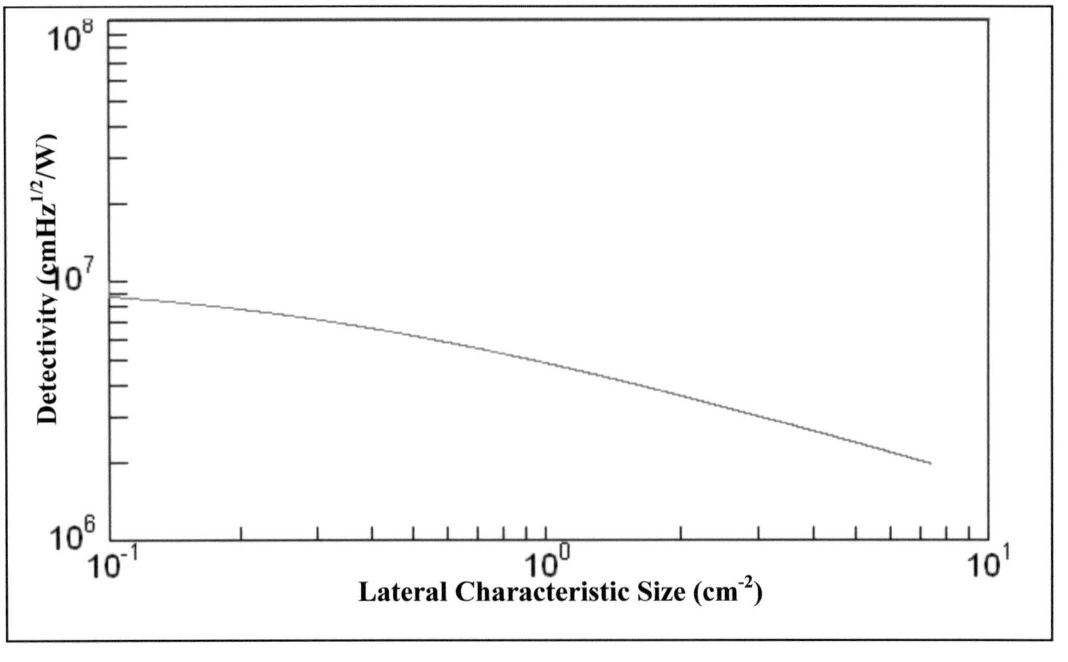

Fig. 3.12. Detectivity against a_{QD}.

2. References

[1] N. J. Withers, K. Sankar, B. A. Akins, T. A. Memon, T. Gu, J. Gu, G. A. Smolyakov, M. R. Greenberg, T. J. Boyle, and M. Osiński, "Rapid degradation of CdSe/ZnS colloidal quantum dots exposed to gamma irradiation", Applied Physics Letters, Vol. 93, 173101, 2008.

[2] S. E. Letant and T.-F. Wang, "Semiconductor quantum dot scintillation under γ-ray irradiation", Nano Letters, Vol. 6, No. 12, pp 2877–2880, 2006.

[3] S. E. Letant, and T.-F. Wang, "Study of porous glass doped with quantum dots or laser dyes under alpha irradiation", Applied Physics Letters, Vol. 88, No. 10, pp. 103110 - 103110-3, 2006.

[4] I. H. Campbell, B. K. Crone, "Quantum-dot/organic semiconductor composites for radiation detection", Advanced Materials, Vol. 18, No. 1, pp. 77-79, 2006. 63

[5] D. S. Burgess, "Quantum dots may enable new radiation detectors", Applied Physics Letters, 103110, 2006.

[6] Serge Luryi, "Impregnated semiconductor scintillator", International Journal of High Speed Electronics and Systems, Vol. 18, No. 4, pp. 973–982, 2008. 36

[7] I. I. Mahmoud, and S. A. Kamel, "Using a simulation technique for switched-mode high voltage power supplies performance study", IEEE Trans. IAS, Vol. 34, No. 5, pp.954-952, 1998.

[8] I. I. Mahmoud, H. A. Konber, and M. S. El_Tokhy, "Block diagram modeling of quantum laser sources", Optics and Laser Technology, Vol. 42, 2010.

[9] M. S. El_Tokhy, "Performance improvement of optical semiconductor sources", Master thesis, Al Azhar university, 2009.

[10] Vissim user's guide, Version 1.2, Visual solutions, Inv., West Ford, MA, 1993. 12

[11] A. N. Immucci, A. Chamson-Reig, R. Z. Stodilka, J. J. L. Carson, K. Yu, D. Wilkinson, and C. Li, "Method for imaging quantum dots during exposure to gamma radiation", Proc. SPIE 7925, Frontiers in Ultrafast Optics: Biomedical, Scientific, and Industrial Applications XI, San Francisco, California, USA, 2011 [doi:10.1117/12.875379].

[12] J. W. Mares, J. Harben, A. V. Thompson, D. W. Schoenfeld, W. V. Schoenfeld, "Gamma radiation induced degradation of operating quantum dot lasers", IEEE Transactions on Nuclear Science, Vol. 55, No. 2, pp. 763– 768, 2008.

[13] V. Ryzhii, I. Khmyrova, M. Ryzhii and V. Mitin, "Comparison of dark current, responsivity and detectivity in different intersubband infrared photodetectors", Semicond. Sci. Technol., Vol. 19, pp. 8–16, 2004.

[14] Imbaby I. Mahmoud, Mohamed S. El_Tokhy, and Hussein A. Konbe, "Model Development of Quantum Dot Devices for c Radiation Detection Using Block Diagram Programming", Technical Brief in Nanotechnology in Engineering and Medicin Journal (NEM), Vol. 2, No. 3, 2011 [DOI: 10.1115/1.4004313].

[15] Chapter 4, "Hypothesis of diffusion-limited growth", http://scholar.lib.vt.edu/theses/available/etd-04262005-81042/unrestricted/Ch4Hypothesis.pdf

[16] V. Ryzhii, I. Khmyrova, V. Pipa, V. Mitin and M. Willander, "Device model for quantum dot infrared photodetectors and their dark-current characteristics", Semiconductor Science and Technology, Vol. 16, 331–338, 2001.

[17] A. Nasr, "Modeling of solid state photodetectors for ionization radiation and optical fiber communications", A doctoral thesis, Al Azhar university, 2003.

[18] Chee Hing Tan, Souye C. Liew Tat Mun, Peter Vines, John P. R. David and Mark Hopkinson, "Measurement of noise and gain in quantum dot infrared photodetectors (QDIPs)", 4th EMRS DTC Technical Conference-Edinburgh, 2007.

[19] P. Bhattacharya, X. H. Su, S. Chakrabarti, G. Ariyawansa and A. G. U. Perera, "Characteristics of a tunneling quantum-dot infrared photodetector operating at room temperature", Applied Physics Letters, Vol. 86, 2005.

[20] P. Martyniuk and A. Rogalski, "Insight into performance of quantum dot infrared photodetectors", Bulletin of the Polish Academy Of Sciences, Technical Sciences, Vol. 57, No. 1, 2009.

[21] Xuejun Lu, Jarrod Vaillancourt and Mark J Meisner, "A modulation-doped longwave infrared quantum dot photodetector with high photoresponsivity", Semicond. Sci. Technol., Vol. 22, pp. 993–996, 2007.

[22] Y. Matsukura, Y. Uchiyama, H. Yamashita, H. Nishino and T. Fujii, "Responsivity-dark current relationship of quantum dot infrared photodetectors (QDIPs)", Infrared Physics and Technology, 2009.

[23] Shahram Mohammad Nejad, Saeed Olyaee and Maryam Pourmahyabadi, "Optimal dark current reduction in quantum well 9 μm GaAs/AlGaAs infrared photodetectors with improved detectivity", American Journal of Applied Sciences, Vol. 5, No. 8, pp. 1071-1078, 2008.

[24] D. G. Nahri, and A. N. Naeimi, "Simulation of static characteristics of self-assembled QD lasers", World Applied Science Journal, Vol. 11, No. 1, 2010.

[25] V. Ryzhii, V. Pipa, I. Khmyrova, V. Mitin, and M. Willander, "Dark current in quantum dot infrared photodetectors", Japan J. Applied Physics, Vol. 39, No. 2000, pp. L1283-L1285, 2000.

[26] P. Martyniuk and A. Rogalski, "Insight into performance of quantum dot infrared photodetectors", Bulletin of the Polish Academy of Sciences, Technical Sciences, Vol. 57, No. 1, 2009.

[27] P. Martyniuk, S. Krishna, and A. Rogalski, "Assessment of quantum dot infrared photodetectors for high temperature Operation", Journal of Applied Physics, Vol. 104, 2008.

[28] N. Li, D.-Y. Xiong, X.-F. Yang, W. Lu, W.-L. Xu, C.-L. Yang, Y. Hou, and Y. Fu, "Dark currents of GaAs/AlGaAs quantum-well infrared photodetectors", Appl. Phys. A, Vol. 89, pp. 701–705, 2007 [doi: 10.1007/s00339-007-4142-2].

[29] A. Nasr, and M. B. El_Mashade, "Theoretical comparison between quantum well and dot infrared photodetectors", Optoelectronics IEE Proceedings, Vol. 153, No. 4, pp. 183-190, 2006 [doi: 10.1049/ip-opt:20050029].

[30] V. Ryzhii, I. Khmyrova, M. Ryzhii, and M. Ershov, "Comparison studies of infrared phototransistors with a quantum-well and a quantum-wire base", Journal De Physique IV 6, pp. C3-157-C3-161, 1996 [doi: org/10.1051/jp4:1996324].

[31] A. Rogalski, "Insight on quantum dot infrared photodetectors", 2nd National Conference on Nanotechnology, Vol. 146, pp. 1-9, 2009.

[32] Mohamed B. El Mashade, M. Ashry and A. Nasr, "Theoretical analysis of quantum dot infrared photodetectors", Semiconductor Science and Technology, Vol. 18, pp. 891–900, 2003.

4

Quantum Dots Semiconductors Obtained by Microwaves Heating

Idalia Gómez
Universidad Autónoma de Nuevo León
México

1. Introduction

The recent widespread interest in semiconductor quantum dots (QDs) is due largely to their distinct optical properties, including broad absorption bands, narrow, size-tunable emission bands, and excellent photostability. Physically, the quantum properties of QDs such as the size-dependent fluorescence emission occur in electron-hole pairs (excitons) that are confined to dimensions that are smaller than the electron-hole distance (exciton diameter). As a result of this condition, the state of free charge carriers within a nanocrystal is quantized and the spacing of the discrete energy states (emission colors) is linked to the nanoparticle size of. The combination of small size, high photostability, and size-tunable emission properties makes quantum dots highly attractive probes for biological, biomedical, and bioanalytical imaging applications.

As an alternative to thermally driven synthesis, microwave heating has been applied successfully to a variety of chemical reactions, including those that impact the fields of materials and nanoscience. For example, microwave techniques have been used to prepare a wide range of materials including nanocrystalline composite solid electrolytes such as Bi_2O_3-HfO_2-Y_2O_3, deep-red-emitting CdSe quantum dots, carbon-coated core shell structured copper and nickel nanoparticles in a ionic liquid, Bi_2Se_3 nanosheets, Au@Ag core-shell nanoparticles. Overall, while microwave assisted synthesis of quantum dots and other nanomaterials have been reported, including ZnSe(S) quantum dots synthesis using microwave heating, in comparison to the strategies reported herein, these procedures often result in QDs of varying morphologies or aggregated crystals with comparably poor luminescent properties, such as broad emission bands.

This chapter presents a brief review of the latest developments in the field of semiconductors quantum dots, with a particular focus on synthesis by microwave heating and their optical properties effect. This paper will illustrate the power of this bottom-up synthetic approach by presenting a few case studies in which ZnS, CdS and CdSe highly luminescence materials have been successfully synthesized based on microwave heating. In the study described herein, we show that microwave heating is indeed a very powerful strategy for the synthesis of high-quality of semiconductors such as ZnS, CdS and CdSe quantum dots with luminescent properties and offers several advantages over traditional thermally driven approaches. In particular, these advantages include the ability to quickly reach reaction temperatures and a straightforward process control, thus making quantum dot materials with quantum yield (QYs) of 70%, accessible to an increased number of research labs.

2. Quantum dots

Quantum Dots are included in nanotechnology term. However is necessary to establishment of their definition in order to understand why nanochemistry is the way for obtaining these nanostructures. When the characteristic dimension of the nanoparticles is sufficiently small and quantum effects are observed, quantum dots are the common term used to describe such nanoparticles. Then in the literature we can found two approaches for the synthesis of nanostructures top-down and bottom-up. Top-down techniques are related with milling or attrition, repeated quenching and lithography, and then all they produce impurities, high crystalline defects, etc. Alternatively, bottom-up approaches are far more popular in the synthesis of quantum dots and many methods have been developed. For example, quantum dots are synthesized by homogeneous nucleation from liquid or vapor, or by heterogeneous nucleation on substrates. Quantum dots can also be prepared by phase segregation through annealing appropriately designed solid materials at elevated temperatures. Quantum dots can be synthesized by confining chemical reactions, nucleation and growth processes in a small space such as micelles. Various synthesis methods or techniques can be grouped into two categories: thermodynamic equilibrium approach and kinetic approach. In the thermodynamic approach, synthesis process consists of (i) generation of supersaturation, (ii) nucleation, and (iii) subsequent growth. In the kinetic approach, formation of quantum dots is achieved by either limiting the amount of precursors available for the growth such as used in molecular beam epitaxy, or confining the process in a limited space such as aerosol synthesis, micelle synthesis and microwave assisted synthesis. In this chapter, the attention will be focused mainly on the synthesis of quantum dots through microwaves heating. Specifically semiconductors quantum dots like as CdS, ZnS and CdSe.

2.1 Quantum dots semiconductors

Semiconductor nanoparticles synthesis has attracted much interest, due to their size-dependent properties and great potential for many applications, especially as nonlinear optical materials. (Spanhel et al., 1987, Henglein et al, 1983, Rossetti et al, 1985, Sun & Riggs, 1999).1-4 Nanoparticles exhibit unique properties owing to quantum size effects and the presence of a large number of unsaturated surface atoms.

Colloidal semiconductor nanocrystals (NCs) are of great interest for fundamental studies (Heath, 1999) 5 and technical applications such as light-emitting devices, (Hikmet, Talapin & Weller, 2003) 6 lasers,7 (Finlayson et al, 2002) and fluorescent labels.8 (Clapp et al, 2004) Because of their size-dependent photoluminescence tunable across the visible spectrum9 (Bawendi, Carroll, Wilson, Brus, 1992). Besides the development of synthesis techniques to prepare samples with narrow size distributions,10,11 (Murray & Bawendi, 1993, Peng & Peng, 2001) much experimental work is devoted to molecular surface modification to improve the luminescence efficiency12,13 (Spanhel et al, 1987, Talapin et al, 2001) and colloidal stability of the particles or to develop a reliable processing chemistry.14,15 (Wang, Li, Chen & Peng, 2002, Aldana, Wang & Peng, 2001)

Semiconductor nanoparticles are expected to exhibit quantum confinement when their size becomes comparable to the 1s-exciton diameter,16-18 (Brus, 1984, Brus & Steigerwald, 1990,

Stucky & McDougall, 1990, Stucky, 1992, Weller, 1993, Siegel, 1993)which results in the appearance of a quantized eigenspectrum and an increase in the energy gap relative to the band gap (Eg) of the bulk solid. Quantum crystals of CdS, a II-VI semiconductor with 6nm exciton diameter and 2.5eV band gap, have been successfully synthesized using many stabilizers in an effort led by Henglein and Brus.19,20 (Henglein, 1982, Brus et al, 1983, 1983) Various surface-capping agents used to stabilize II-VI semiconductor nanoparticles include polyphosphate,21 (Euchmüller, 2000) trioctylphosphine/trioctylphosphine oxide, and thiols.22 (Vossmeyer et al, 1994)

Synthesis of nanocrystals with a size23,24 (Peng, Wickham & Alivisatos, 1998, O'Brien, Brus & Murray, 2001 and shape25-31 (Trentler et al, 1995, Peng et al, 2000, Pacholski, Kornowski & Weller, 2002, Tang, Kotov & Giersig, 2002, Peng, & Peng, 2002, Lee, Cho & Cheon, 2003, Cho et al, 2005) defined has advance in recent years, to which high temperature approaches (roughly 250-350 °C) in organic solvents, either through organometallic schemes 10,26,32 (Murray, B. N. & Bawendi, 1993, Peng et al, 2000, Cho et al, 2001) or alternative approaches (or greener approaches),33,34 (Sun & Zeng, 2002, Jana, Chen & Peng, 2004),have played a key role and often been regarded as the mainstream synthetic chemistry in the field. Emphasis on synthetic chemistry of nanocrystals is a currently moving into nano-objects with complex structures and compositions,35 (Peng & Thessing, 2005)and formation of three-dimensional (3D) colloidal nanocrystals is especially under development.

Nanocrystals with complex 3D structures are interesting for solar cells, catalysis, sensing, and other surface/shape related properties and applications.36,37 (Gur et al, 2005, Pinna et al, 2004) For instance, CdTe and other semiconductor tetrapods38,39 (Manna et al, 2003, Yu et al, 2003) are ideal structures for fabrication of high performance solar cells 39 (Yu et al, 2003). Such tetrapods, however, were typically formed by a traditional path, atom by atom growing from nucleis, and the intrinsic crystal structures seem to play a key role, i.e., nanolitography. Thus, is not clear how to extend the synthetic methods to different structures. Some reports indicate that nanodots and nanorods can self-assembling into different complex shaped particles.40-42 (Huynh et al, 2002, Chen et al, 2005, Liu & Zeng, 2004). Such complex structures, however, were often quite large, fragile, and/or polycrystalline. Some other reports indicated possibilities of formation of complex nanostructures through 3D attachment.43 (Zitoun et al, 2005) Thus, a general pathway to reach 3D oriented attachment has not yet been achieved.

The recent widespread interest in semiconductor quantum dots (QDs) is due largely to their distinct optical properties, including broad absorption bands, narrow, size-tunable, emission bands, and excellent photostabilities.44 (Bawendi et al, 1990) Physically, the quantum properties of QDs (a size-dependent fluorescence emission) occur in electron-hole pairs (excitons) that are confined to dimensions that are smaller than the electron-hole distance (exciton diameter).44-46 (Bawendi et al, 1990, Brus, 1983, 1986) As the result of this condition, the state of free charge carriers within a nanocrystal is quantized and the spacing of the discrete energy states (emission colors) is linked to their size of nanoparticle. The combination of small size, high photostability, and size-tunable emission properties makes quantum dots highly attractive probes for biological, biomedical, and bioanalytical imaging applications.47-50 (Chan et al, 2002, Parak et al, 2003, Bailey et al, 2004, Murcia & Naumann, 2005)

As an alternative to thermally driven syntheses, microwave heating has been applied successfully to a variety of chemical reactions, including those that impact the fields of materials and nanoscience. For example, microwave techniques have been used to prepare a wide range of materials including nanocrystalline composite solid electrolyte like Bi_2O_3-HfO_2-Y_2O_3 by microwave plasma,51 (Zhen et al, 2006) deep-red-emitting CdSe quantum dots,52 (Wang & Seo, 2006) carbon-coated core shell structured copper and nickel nanoparticles in a ionic liquid,53 (Jacob et al, 2006) Bi_2Se_3 nanosheets,54 (Jiang et al, 2006) Au@Ag core-shell nanoparticles.55 (Tsuji et al, 2006) Overall, while microwave assisted syntheses of quantum dots and other nanomaterials have been reported,56,57 (He et al, 2006a; 2006b) including ZnSe(S) quantum dots synthesis using microwave heating,58 (Jacob et al, 2006) in comparison to the strategies reported herein, these procedures often result in QDs of varying morphologies or aggregated crystals with comparably poor luminescent properties, such as broad emission bands. Indeed, a recent review that surveyed the latest productive routes to high-quality ZnS quantum dots did not include microwave heating 59 (Huifeng et al, 2006). Panda et al, synthesized ZnS rods by microwave heating 60 (Panda et al, 2006), but they do not studied the changes in the band gap absorption according to the particle size, the ZnS nanorods produced by them are not highly luminescent, indeed, they used the same method to prepared nano semiconductor. While they used cycles of microwave heating, we use an uninterrupted process of heating.

In the study described herein, we show that microwave heating can indeed provide a very powerful strategy for the synthesis of high-quality ZnS, CdS and CdSe quantum dots with highly luminescent properties and offers several advantages over traditional thermally driven approaches. In particular, these advantages include get quickly reactions temperatures and a straightforward process control, thus making quantum dot materials with quantum yield (QYs) of 70%, accessible to an increased number of research labs.

3. Experimental procedures

3.1 Materials

Thioacetamide (CH_3CSNH_2) was purchased from MERCK Gehalt (99.0%). $ZnSO_4$ and $CdSO_4$ were obtained from Spectrum Quality Products, Inc. (99.9%), KOH was purchased from MERCK (99.0%). All chemicals were used without additional purification. All solutions were prepared using Milli-Q water (Millipore) as the solvent.

3.2 Preparation of QD´s

3.2.1 ZnS and CdS

Nearly monodisperse ZnS and CdS NPs were obtained by microwave irradiation. The ZnS or CdS NPs solution was prepared by adding freshly prepared either $ZnSO_4$ or $CdSO_4$ solution to a thioacetamide solution at pH 8 in the presence of sodium citrate solution used as stabilizer. The precursors concentration were [S] = $3x10^{-2}$M, [S] = $6x10^{-2}$M, [S] = $8x10^{-2}$M and also using concentrations of precursor [Zn-Cd] = $3x10^{-2}$M. The NPs were prepared under microwave irradiation for 1 min at 905W of power. The NPs samples were taken when temperature decreases until ambient temperature for further analysis.

3.2.2 CdSe

Nearly monodisperse CdSe was obtained by microwave irradiation. In this case were used Cd and Na2SeSO3 as precursors in the presence of sodium citrate solution used as stabilizer. The concentrations of precursor [Cd] = 3x10-2M. The experiments were doing at different pH values in order to observe the blue shift related with quantum confinement as quantum dot.

3.3 Apparatus

The microwave system used for the synthesis of NPs operates at 1150W, 2.45 GHz, working at 90% of power under continuous heating. UV-vis absorption spectra were obtained using a Perkin Elmer UV-vis Lambda 12 spectrophotometer. For luminescence quantum yield measurements, a dilute solution of coumarin 1 in ethanol was used as standard. Both the nanoparticle dispersion and the coumarin 1/ethanol solution were adjusted to have an absorbance of 0.10. A corrected luminescence integrated area was used to calculate the quantum yield. Fluorescence experiments were performed using a Perkin Elmer PL Lambda 12 spectrofluorimeter using a wavelength of excitation of 250nm. All optical measurements were performed at room temperature under ambient conditions. Samples were precipitated with ethanol and dried in a vacuum oven for XRD characterization. The XRD patterns were obtained from a Siemmens D5000 Cu Kα (λ = 1.5418Å) diffractometer. AFM images were recorded in a Quesant Q-Scope 3500 atomic force microscope using contact mode. HRSEM was used.

4. Results and discussions

4.1 CdS and ZnS

4.1.1 Uv-Vis

Figure 1, displays the UV-vis absorption of ZnS, CdS NPs synthesized trough microwave irradiation at 905W. A blue-shift was observed in the absorption of ZnS and CdS NPs. This is indicative of the NPs formation and, the blue shift is due to the concentration decreasing of S2- ions in samples of ZnS but, a contraries effect in samples of CdS was observed. The size of the ZnS NPs increased when the concentration of S2- also increases, which results in a gradual red-shift. This kind of absorption provides evidence of the quantum size effect and the presence of particles of nanometer size. Different effect was observed in samples of CdS the increase of S2- ions produce the blue shift, theoretical works have shown that the absorption threshold provides a reasonable estimation of the particle size as function of the position and spacing of the threshold absorption.59 (Huifeng et al, 2006) The blue-shift indicates the size of the particle can be changed by different addition of S2- ions and produce a variation of the optical properties of the ZnS and CdS NPs.

The most striking features are as follows: first one all the sodium citrate stabilizer particles show a well-developed behavior of classical semiconductor at nanometer scale, and a well-developed curve near the onset of absorption which is ascribed to the first excitonic (1s-1s) transition. In some cases even higher energy transitions are observed. Second one with decreasing particle size the transition energy shift to higher values as a consequence of the size quantization effect.

It has been reported that ZnS and CdS are direct band gap semiconductors and therefore plots of (Ahv)2 versus hv should be straight lines with intercepts on the energy axis giving the band gaps of the NPs.60 (Panda et al, 2006) The band gap of the ZnS nanoparticles using the three different concentrations of [S] was found to be 4.32eV, but for CdS the values for Eg were 2.8, 2.74 and 2.61eV according to concentration of [S] 3x10-2, 6x10-2 and 8x10-2M respectively, these are according to color effect observed to CdS nanoparticles.

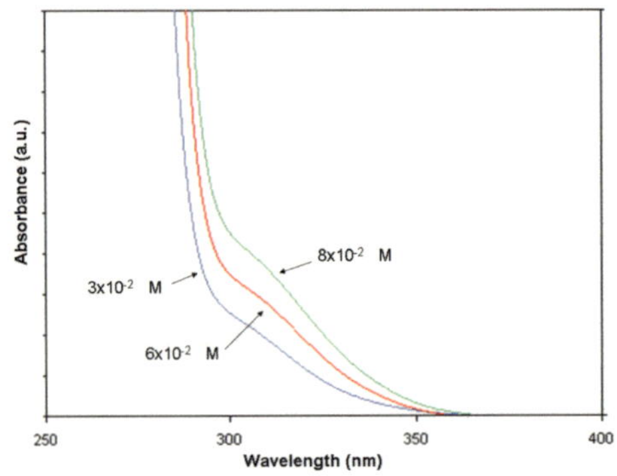

A

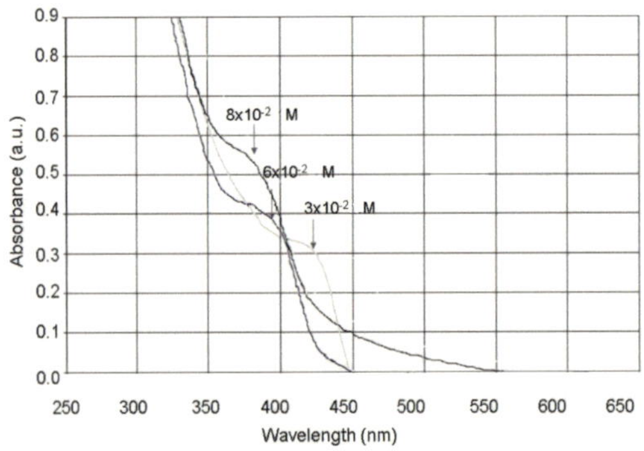

B

Fig. 1. UV-Vis spectra for A)ZnS and B)CdS.

4.1.2 PL

Figure 2 shows the room temperature photoluminescence spectrum of samples of NPs. The line width on emission is shifted to the red with the reduction of the particle size. This shift is the result of a combination of relaxation into shallow trap states and the size distribution. No deep trap emission features suggest highly monodisperse samples; this was observed principally for ZnS samples. High quantum yields and narrow emission line widths indicate growth of NPs with few electronic defect sites. The sharp luminescence is a dramatic example of the efficiency of the capping stabilizer in electronically passivity the crystallites from chemical degradation yielding robust systems. Samples stored in the original growth solution still show strong, sharp emission after storage for more than a month. A luminescence quantum yield of 70% and 60% was measured for the ZnS and CdS NPs respectively using a dilute solution of Coumarin 1 in ethanol as standard.

These room temperature optical experiments, carried out on common laboratory equipment, demonstrate the benefits of high quality samples and point to the potential of more sophisticated optical studies on these samples. Under UV light the NPs shows highly luminescent as is shown in figure 3. To see this effect each sample must be irradiated under UV light using a wavelength with energy higher than the band gap found for each sample.

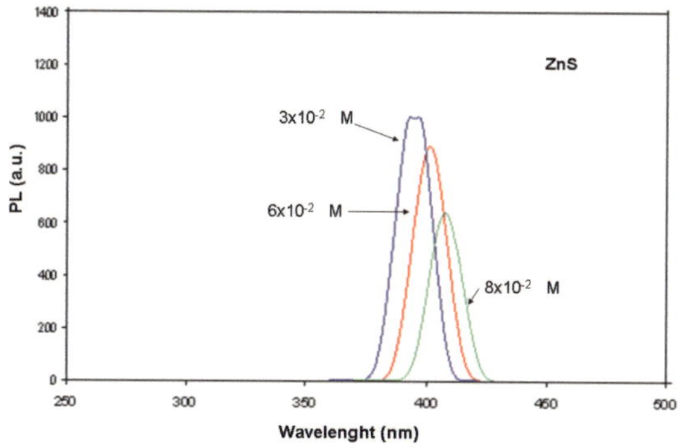

A

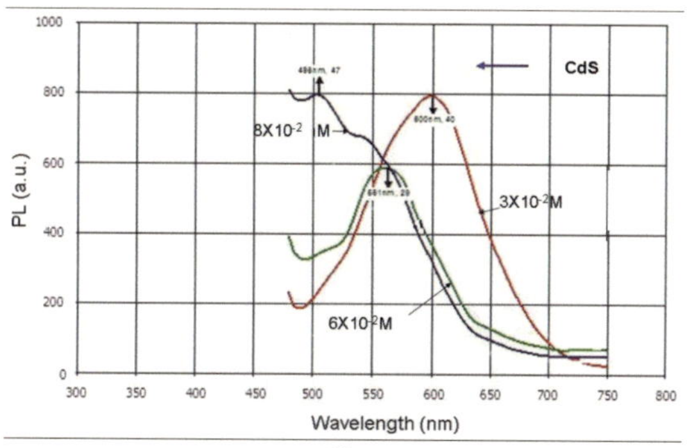

B

Fig. 2. PL spectra for ZnS and CdS.

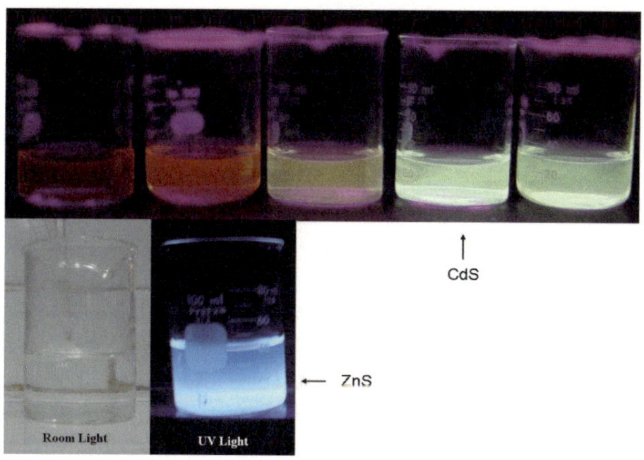

Fig. 3. Photos for ZnS and CdS under UV radiation.

4.1.3 X-ray diffraction

The as-prepared chalcogenide were characterized by X-ray powder diffraction, which showed a match with the diffraction pattern published in the literature. All the diffraction peaks can be indexed to the hexagonally structured for ZnS. The broad nature of the ZnS and CdS XRD peaks shows that the sizes of the quantum dots are very small. The broad diffraction peak at 28.6° (2θ scale) arising from (111) reflections from ZnS is assigned to the cubic sphalerite form (JCPDS card, file no. 50566). The obviously weaker diffractions at 47.50, 56.30 (220) and (311) reflections are of the cubic sphalerite ZnS form, respectively. Figure 4 A) present a diffraction spectrum for the sample obtained at concentration of [S] 8x10-2M. The CdS synthesized is present in wurtzite form (see Figure 4 B).

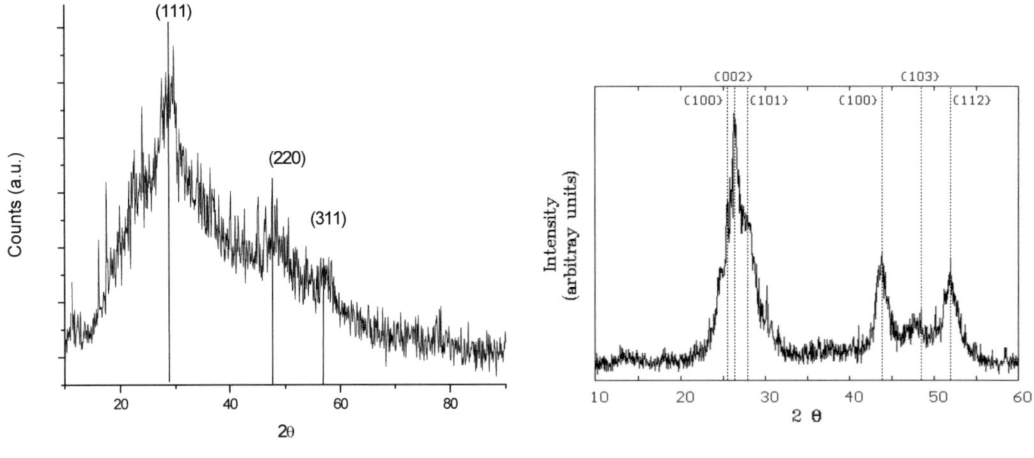

A B

Fig. 4. Diffraction patterns of synthesized A)ZnS and B)CdS.

4.1.4 AFM, HRSEM and TEM

In AFM analysis for ZnS samples was observed the formation of islands (Figure 5A), in some of this we observe the formation of centers in the middle of the islands with a height about 20-40nm, this kind of morphology is not present in all cases, closer to this islands is possible to observe particles of minor size (see Figure 5B). HRSEM analysis in the same samples, ZnS, was observed the formation of singular agglomerates of nanoparticles with sizes minor to 5nm, with figures like to the observed by AFM. The samples of CdS shown agglomerates with minor sizes. Figure 6 shows two images representative to this phenomenon. Figure 7 show TEM micrographs for ZnS and CdS and can see nanoparticles with sizes already of 6 nm.

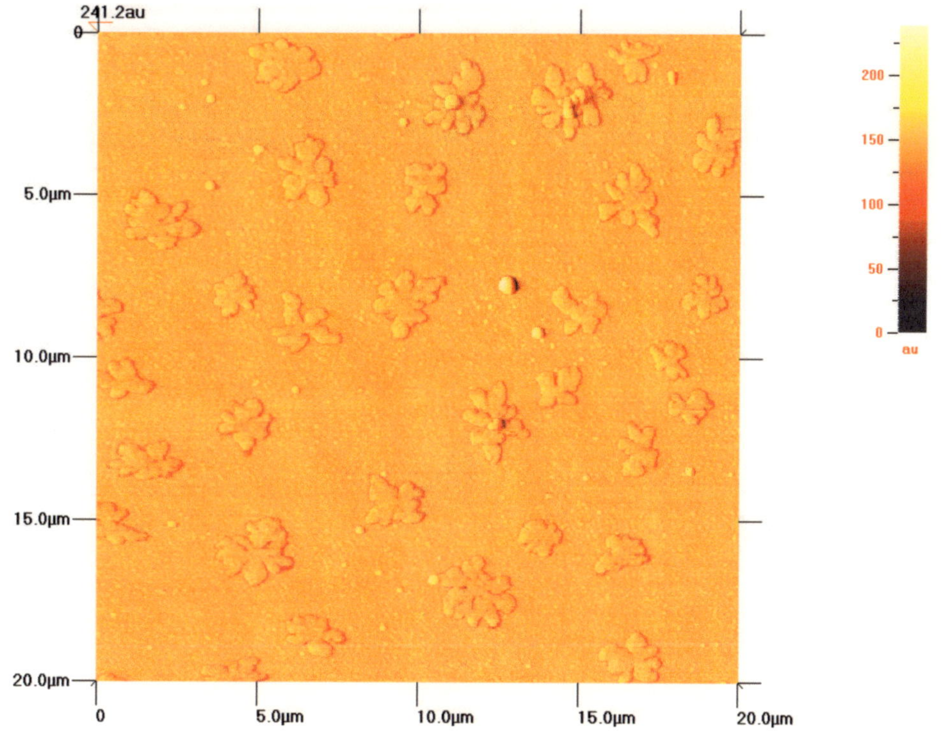

A

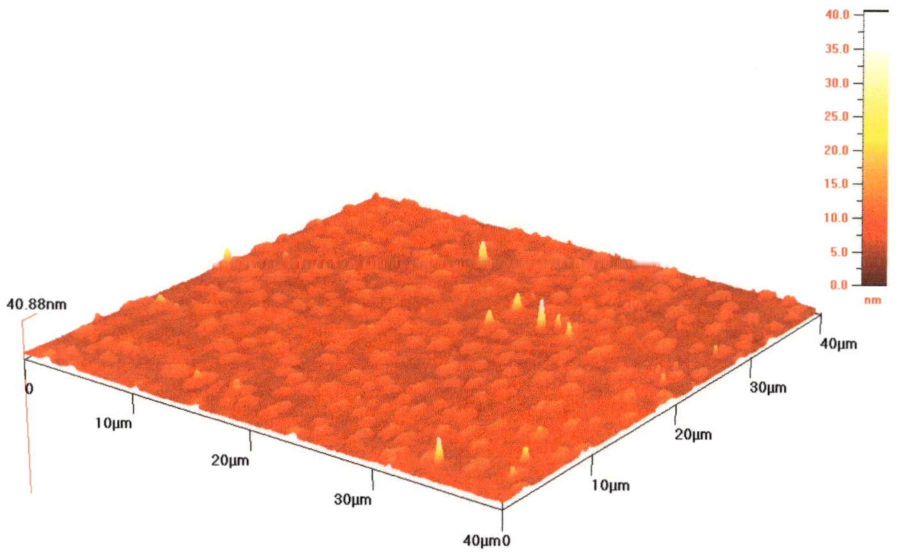

B

Fig. 5. AFM images of ZnS samples synthesized by microwave heating.

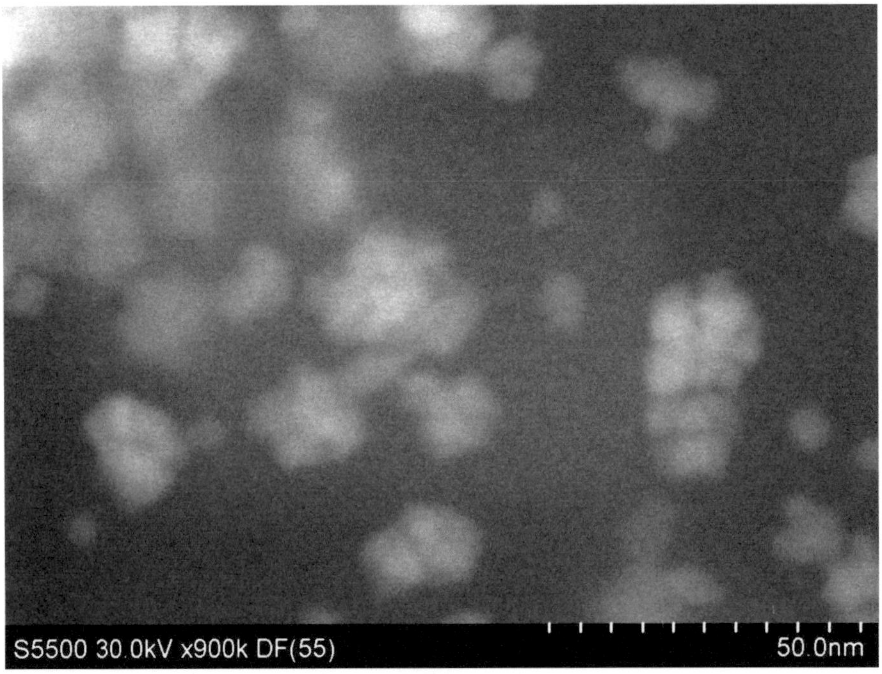

A

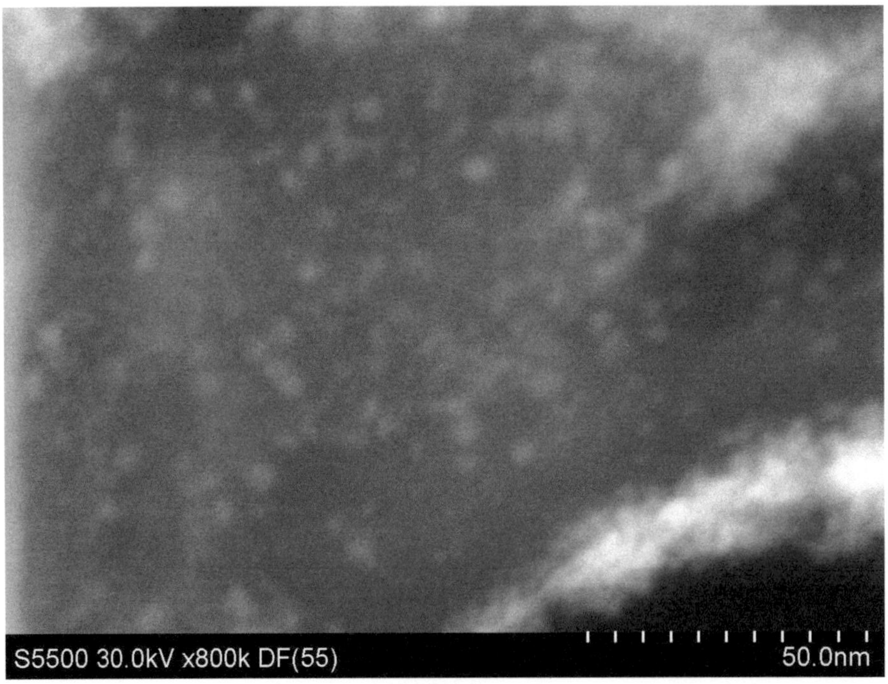

B

Fig. 6. HRSEM images of ZnS and CdS samples, A and B respectively.

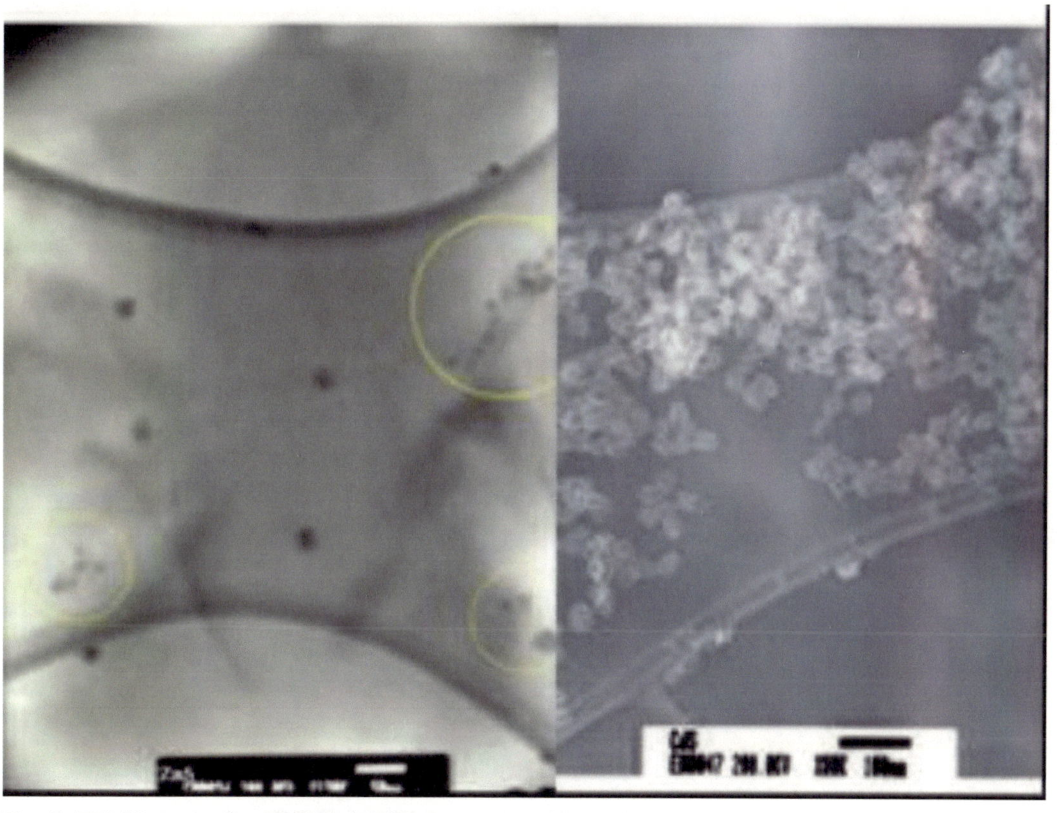

Fig. 7. TEM images for ZnS and CdS.

It is interesting to note the similitude between the results observed by different techniques of characterization. The sizes observed by TEM report sizes already of 6 nm, that are identified as quantum dots.

4.2 CdSe

Figure 8, displays the UV-vis absorption of CdSe NPs synthesized trough microwave irradiation at 905W. A blue-shift was observed in the absorption of CdSe NPs at pH values from 4 to 9. This is indicative of the NPs formation and provides evidence of the quantum size effect and the presence of particles of nanometer size. Different effect was observed when the pH value is bigger than 9. This effect is due an accelerated growth caused by particles joining together due to electrostatic forces in samples of CdSe as can seen in the figure 9 that shown a FE-SEM image with agglomerates at 50nm and higher. The increase of pH produce the red shift, theoretical works have shown that the absorption threshold provides a reasonable estimation of the particle size as function of the position and spacing of the threshold absorption.59 (Huifeng et al, 2006) The blue-shift indicates the size of the particle can be changed by different pH values and produce a variation of the optical properties of the CdSe NPs. Figure 10 present a PL spectra for the sample of CdSe prepared at 6 pH value. A luminescence quantum yield of 85% was measured for the CdSe NPs respectively using a dilute solution of Coumarin 1 in ethanol as standard.

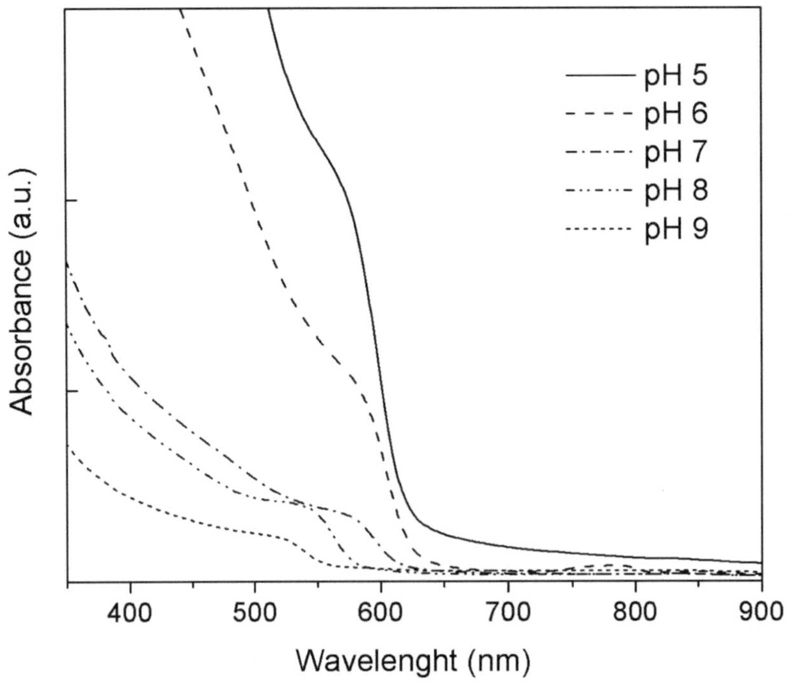

Fig. 8. UV-Vis spectra for CdSe at different pH values.

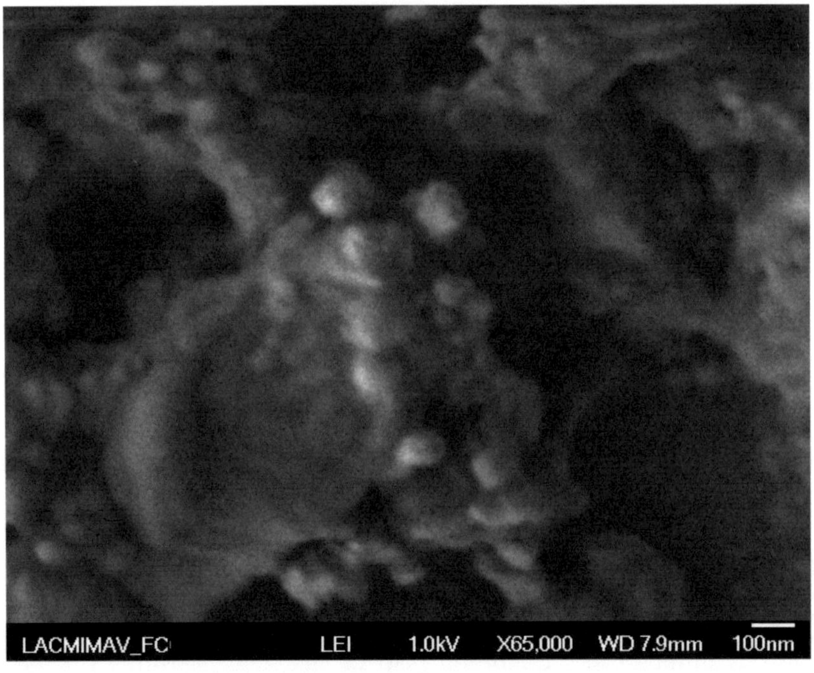

Fig. 9. FE-SEM image for CdSe sample prepared at 9 pH.

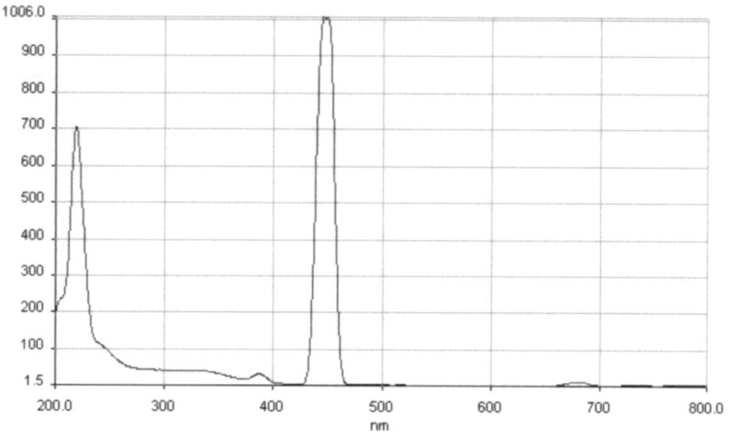

Fig. 10. PL for CdSe.

5. Conclusions

UV-vis spectra of the nanoparticles synthesized show a blue-shift due to the quantum confinement and the reduction of the particle size. This effect is due the increase of the concentration of metal ion. The band gap of the semiconductors increases when the particle size decreases. The crystal structure of the ZnS synthesized is cubic sphalerite form and for CdS was cubic type zinc blend. The morphology of the prepared ZnS shows islands with a nanocenter as well as nanoparticles of about 100 nm. The ZnS NPs obtained shown high monodispersity according to PL analysis. The high luminescence is present when the NPs are irradiated with UV light using energy higher than the band gap value found for each sample, this property is due to the reduction of the particle at nanometer scale. HRSEM and TEM analyses showed nanoparticles of 5nm and nanostructures of agglomerates with sizes already of 10-20 nm. For the samples of CdSe by UV-Vis was observed the effect of quantum confinement at pH 5 and agglomeration effect was observed by FE-SEM according the increase of pH. Synthesis by microwave heating provides a very powerful option to prepare ZnS, CdS and CdSe nanoparticles with highly luminescent properties.

6. Acknowledgements

This document contains part of different works developed during the thesis of students of posgrade and Bachelor Science at Chemistry Science Faculty of the UANL, México, under the direction of the author and sponsored by the Mexican Council for Science and Technology (CONACYT), and the University itself through its Aid Program to Scientific and Technological Research (PAICYT).

7. References

[1] Spanhel, L.; Hoasse, M.; Weller, H. J.; Henglein, A. Photochemistry of colloidal semiconductors. 20. Surface modification and stability of strong luminescing CdS particles *J. Am. Chem. Soc.* 1987, *109*, 5649. ISSN 0002-7863

[2] Henglein, A.; Gutierrez, M. Ber. Bunsen-Ges. *Phys. Chem.* 1983, *87*, 852.

[3] Rossetti, R.; Hull. R.; Gibson, J. M.; Brus, L. E. *J. Chem. Phys.* 1985, *82*, 552. Excited electronic states and optical spectra of ZnS and CdS crystallites in the 15 to 50 Å size range: Evolution from molecular to bulk semiconducting properties ISSN 0021-9606

[4] Sun, Y.; Riggs, J. E. *Int. Rev. Phys. Chem.* 1999, *18*, 43. Organic and inorganic optical limiting materials. From fullerenes to nanoparticles ISSN 0144-235X

[5] Heath, J. R. *Acc. Chem. Res.* 1999, *32*, 389-414. Spectroscopy of Single CdSe Nanocrystallites, Stephen Empedocles and Moungi Bawendi, Acc. Chem. Res., 1999, 32 (5), pp 389–396

[6] Hikmet, R. A. M.; Talapin, D. V.; Weller, H. *J. Appl. Phys.* 2003, *93*, 3509-3514.

[7] Finlayson, C. E.; Russell, D. M.; Ramsdale, C. M.; Ginger, D. S.; Silva, C.; Greenham, N. C. *Adv. Funct. Mater.* 2002, *12*, 537-540.

[8] Clapp, A. R.; Medintz, I. L.; Mauro, J. M.; Fisher, B. R.; Bawendi, M. G.; Mattoussi, H. *J. Am. Chem. Soc.* 2004, *126*, 301-310. Fluorescence Resonance Energy Transfer Between Quantum Dot Donors and Dye-Labeled Protein Acceptors.

[9] Bawendi, M. G.; Carroll, P. J.; Wilson, W. L.; Brus, L. E. *J. Chem. Phys.* 1992, *96*, 946-954.

[10] Murray, C. B. N., D. J.; Bawendi, M. G. *J. Am. Chem. Soc.* 1993, *115*, 8706-8715. Synthesis and characterization of nearly monodisperse CdE (E = sulfur, selenium, tellurium) semiconductor nanocrystallites

[11] Peng, Z. A.; Peng, X. G. *J. Am. Chem. Soc.* 2001, *123*, 183-184. Formation of High-Quality CdTe, CdSe, and CdS Nanocrystals Using CdO as Precursor

[12] Spanhel, L.; Haase, M.; Weller, H.; Henglein, A. *J. Am. Chem. Soc.* 1987, *109*, 5649-5655. Photochemistry of colloidal semiconductors. 20. Surface modification and stability of strong luminescing CdS particles

[13] Talapin, D. V.; Rogach, A. L.; Kornowski, A.; Haase, M.; Weller, H. *Nano Lett.* 2001, *1*, 207-211. Highly Luminescent Monodisperse CdSe and CdSe/ZnS Nanocrystals Synthesized in a Hexadecylamine−Trioctylphosphine Oxide−Trioctylphospine Mixture.

[14] Wang, Y. A.; Li, J. J.; Chen, H. Y.; Peng, X. G. *J. Am. Chem. Soc.* 2002, *124*, 2293-2298. Stabilization of Inorganic Nanocrystals by Organic Dendrons

[15] Aldana, J.; Wang, Y. A.; Peng, X. G. *J. Am. Chem. Soc.* 2001, *123*, 8844-8850. Photochemical Instability of CdSe Nanocrystals Coated by Hydrophilic Thiols

[16] Brus, L. E. *J. Chem. Phys.* 1984, *80*, 4403. Electron–electron and electron-hole interactions in small semiconductor crystallites: The size dependence of the lowest excited electronic state

[17] Brus, L. E.; Steigerwald, M. L. *Acc. Chem. Res.* 1990, *23*, 183. Semiconductor crystallites: a class of large molecules (b) Stucky, G. D.; McDougall, J. E. *Science* 1990, *247*, 669. (c) Stucky, G. D. *Prog. Inorg. Chem.* 1992, *40*, 99.

[18] Weller, H. *Adv. Mater.* 1993, *5*, 88. (b) Siegel, R. W. *Phys. Today* 1993, *46* (10), 64.

[19] Henglein, A. *J. Phys. Chem.* 1982, *86*, 2291. Photochemistry of colloidal cadmium sulfide. 2. Effects of adsorbed methyl viologen and of colloidal platinum

[20] Rossetti, R.; Nakahara, S.; Brus, L. E. *J. Chem. Phys.* 1983, *79*, 1086. Quantum size effects in the redox potentials, resonance Raman spectra, and electronic spectra of CdS crystallites in aqueous solution (b) Brus, L. E. *J. Phys. Chem.* 1983, *79*, 5566.

[21] Euchmüller, A. *J. Phys. Chem. B* 2000, *104*, 6514 and references therein. Structure and Photophysics of Semiconductor Nanocrystals

[22] Vossmeyer, T.; Katsikas, L.; Giersig, M.; Popovic, I. G.; Diesner, K.; Chemseddine, A.; Euchmu¨ ller, A.; Weller, H. *J. Phys. Chem.* 1994, *98*, 7665. ISSN 1520-6106

[23] Peng, X.; Wickham, J.; Alivisatos, A. P. *J. Am. Chem. Soc.* 1998, *120* (21), 5343-5344. Kinetics of II-VI and III-V Colloidal Semiconductor Nanocrystal Growth: "Focusing" of Size Distributions

[24] O'Brien, S.; Brus, L.; Murray, C. B. *J. Am. Chem. Soc.* 2001, *123* (48), 12085-12086. Synthesis of Monodisperse Nanoparticles of Barium Titanate: Toward a Generalized Strategy of Oxide Nanoparticle Synthesis

[25] Trentler, T. J.; Hickman, K. M.; Goel, S. C.; Viano, A. M.; Gibbons, P. C.; Buhro, W. E. *Science* 1995, *270* (5243), 1791-4.

[26] Peng, X.; Manna, U.; Yang, W.; Wickham, J.; Scher, E.; Kadavanich, A.; Alivisatos, A. P. *Nature* 2000, *404* (6773), 59-61.

[27] Pacholski, C.; Kornowski, A.; Weller, H. *Angew. Chem., Int. Ed.* 2002, *41* (7), 1188-1191.

[28] Tang, Z.; Kotov, N. A.; Giersig, M. *Science* 2002, *297* (5579), 237-240.

[29] Peng, Z. A.; Peng, X. *J. Am. Chem. Soc.* 2002, *124* (13), 3343-3353. Nearly Monodisperse and Shape-Controlled CdSe Nanocrystals via Alternative Routes: Nucleation and Growth

[30] Lee, S.-M.; Cho, S.-N.; Cheon, *J. Adv. Mater.* 2003, *15* (5), 441-444.

[31] Cho, K.-S.; Talapin, D. V.; Gaschler, W.; Murray, C. B. *J. Am. Chem. Soc.* 2005, *127* (19), 7140-7147. Designing PbSe Nanowires and Nanorings through Oriented Attachment of Nanoparticles

[32] Cho, K.; Koh, H.; Park, J.; Oh, S. J.; Kim, H.-D.; Han, M.; Park, J. H.; Chen, C. T.; Kim, Y. D.; Kim, J. S.; Jonker, B. T. *Phys. Rev. B: Condens. Matter Mater. Phys.* 2001, *63* (15), 155203/1-155203/7.

[33] Sun, S.; Zeng, H. *J. Am. Chem. Soc.* 2002, *124* (28), 8204-8205. Size-Controlled Synthesis of Magnetite Nanoparticles

[34] Jana, N. R.; Chen, Y.; Peng, X. *Chem. Mater.* 2004, *16* (20), 3931-3935. Size- and Shape-Controlled Magnetic (Cr, Mn, Fe, Co, Ni) Oxide Nanocrystals via a Simple and General Approach ISSN 0897-4756

[35] Peng, X.; Thessing, J. Struct. Bonding 2005, 118 (Semiconductor Nanocrystals and Silicate Nanoparticles), 79-119.

[36] Gur, I.; Fromer, N. A.; Geier, M. L.; Alivisatos, A. P. *Science* 2005, *310* (5747), 462-465.

[37] Pinna, N.; Neri, G.; Antonietti, M.; Niederberger, M. *Angew. Chem., Int. Ed.* 2004, *43* (33), 4345-4349.

[38] Manna, L.; Milliron, D. J.; Meisel, A.; Scher, E. C.; Alivisatos, A. P. *Nat. Mater.* 2003, *2* (6), 382-385.

[39] Yu, W. W.; Wang, Y. A.; Peng, X. *Chem. Mater.* 2003, *15* (40), 4300- 4308. Formation and Stability of Size-, Shape-, and Structure-Controlled CdTe Nanocrystals: Ligand Effects on Monomers and Nanocrystals ISSN 0897-4756

[40] Huynh, W. U.; Dittmer, J. J.; Alivisatos, A. P. *Science* 2002, *295* (5564), 2425-2427.

[41] Chen, J.; Herricks, T.; Xia, Y. *Angew. Chem., Int. Ed.* 2005, *44* (17), 2589- 2592.

[42] Liu, B.; Zeng, H. C. *J. Am. Chem. Soc.* 2004, *126* (26), 8124-8125. Mesoscale Organization of CuO Nanoribbons: Formation of "Dandelions"

[43] Zitoun, D.; Pinna, N.; Frolet, N.; Belin, C. *J. Am. Chem. Soc.* 2005, *127* (43), 15034-15035. Single Crystal Manganese Oxide Multipods by Oriented Attachment

[44] Bawendi, M. G.; Steigerwald, M. W.; Brus, L. E. *Annu. Rev. Phys. Chem.* 1990, *41*, 477.

[45] Brus, L. E., J. *Chem. Phys.* 1983, *79*, 5566-5571. A simple model for the ionization potential, electron affinity, and aqueous redox potentials of small semiconductor crystallites ISSN 0021-9606

[46] Brus, L. E., J. *Chem. Phys.* 1986, *90*, 2555-2570. Time dependent calculations of the absorption spectrum of a photodissociating system with two interacting excited electronic states ISSN 0021-9606

[47] Chan, W. C. W.; Maxwell, D. J.; Gao, X.; Bailey, R. E.; Han; M.; Nie, S. *Curr. Opin. Biotechnol.* 2002, *13*, 40.

[48] Parak, W. J.; Gerion, D.; Pellegrino, T.; Znchet, D.; Micheel, C.; Williams, S. C.; Boudreau, R.; Le Gros, M. A.; Larabell, C. A.; Alivisatos, A. P. *Nanotechnology* 2003, *14*, R15.

[49] Bailey, R. E.; Smith, A. M.; Nie, S. *Physica E* 2004, *25*, 1., Quantum dots in biology and medicine, Physica E: Low-dimensional Systems and Nanostructures, Volume 25, Issue 1, October 2004, Pages 1-12, ISSN 1386-9477,

[50] Murcia, M.; Naumann, C. A. In *Biofunctionalization of Nanomaterials*; Wiley: Weinheim, 2005; p 1.

[51] Zhen, Q.; Kale, G. M.; He, W.; Liu, J., Chem. Mater., 2007, 19 (2), pp 203–210; Microwave Plasma Sintered Nanocrystalline Bi2O3−HfO2−Y2O3 Composite Solid Electrolyte, ISSN 0897-4756

[52] Wang, Q.; Seo, D.-K. *Chem. Mater.;* 2006; *18* (24); 5764-5767. Synthesis of Deep-Red-Emitting CdSe Quantum Dots and General Non-Inverse-Square Behavior of Quantum Confinement in CdSe Quantum Dots ISSN 0897-4756

[53] Jacob, D. S.; Genish, I.; Klein, L.; Gedanken, A. *J. Phys. Chem. B.;* 2006; *110* (36); 17711-17714. Carbon-Coated Core Shell Structured Copper and Nickel Nanoparticles Synthesized in an Ionic Liquid. ISSN 1520-6106

[54] Jiang, Y.; Zhu, Y.-J.; Cheng, G.-F. *Cryst. Growth Des.;* 2006; *6* (9); 2174-2176. Synthesis of Bi2Se3 Nanosheets by Microwave Heating Using an Ionic Liquid ISSN 1528-7483

[55] Tsuji, M.; Miyamae, N.; Lim, S.; Kimura, K.; Zhang, X.; Hikino, S.; Nishio, M. *Cryst. Growth Des.;* 2006; *6* (8); 1801-1807. Crystal Structures and Growth Mechanisms of Au@Ag Core−Shell Nanoparticles Prepared by the Microwave−Polyol Method ISSN 1528-7483

[56] He, Y.; Lu, H.-T.; Sai, L.-M.; Lai, W.-Y.; Fan, Q.-L.; Wang, L.-H.; Huang, W. *J. Phys. Chem. B.;* 2006; *110* (27); 13370-13374. Microwave-Assisted Growth and Characterization of Water-Dispersed CdTe/CdS Core−Shell Nanocrystals with High Photoluminescence ISSN 1520-6106

[57] He, Y.; Lu, H.-T.; Sai, L.-M.; Lai, W.-Y.; Fan, Q.-L.; Wang, L.-H.; Huang, W. *J. Phys. Chem. B.;* 2006; *110* (27); 13352-13356. Synthesis of CdTe Nanocrystals through Program Process of Microwave Irradiation, ISSN 1520-6106

[58] Jacob, D. S.; Bitton, L.; Grinblat, J.; Felner, I.; Koltypin, Y.; Gedanken, A. *Chem. Mater.;* 2006; *18*(13); 3162-3168. Are Ionic Liquids Really a Boon for the Synthesis of Inorganic Materials? A General Method for the Fabrication of Nanosized Metal Fluorides ISSN 0897-4756

[59] Huifeng Qian, Xin Qiu, Liang Li, and Jicun Ren. *J. Phys. Chem. B*; 2006; *110*(18); 9034–9040. Microwave-Assisted Aqueous Synthesis: A Rapid Approach to Prepare Highly Luminescent ZnSe(S) Alloyed Quantum Dots, ISSN 1520-6106

[60] Panda, A. B.; Glaspell, G.; El-Shall, M. S. *J. Am. Chem. Soc.;* 2006; *128*(9); 2790-2791. Microwave Synthesis of Highly Aligned Ultra Narrow Semiconductor Rods and Wires

5

Self-Assembled Nanodot Fabrication by Using PS-PDMS Block Copolymer

Miftakhul Huda, You Yin and Sumio Hosaka
Graduate School of Engineering, Gunma University
Japan

1. Introduction

The downsizing of nanolithography technology has given great benefit on achieving faster, low power consumption, and high integrated structure of electronics devices. Therefore, this nanolithography technology has drawn many scientists and engineers to be involved and put their main goal on this field in many decades. To obtain nanostructures especially nanodot, there are many methods which have been developed. Before block copolymer self-assembly technique is presented, those methods to fabricate nanostructures are discussed. Those methods could be divided into 2 large categories. Those are top-down method and bottom-up methods. The methods to fabricate nanodot also could be divided into top-down and bottom-up methods. Top-down method patterns material at large scale by reducing its dimension to the nanoscale. Bottom-up methods arrange atoms or molecules to form nanostructures.

Top-down methods could be defined as patterning by scraping material into smaller dimension. The main representatives of top-down methods are lithography, Electron Beam (EB) Drawing, and Ion Beam technique. Many experiments in fabrication nanodots using these techniques have been reported.

In lithography technique, a beam of light (typically a UV light) passes through the mask and a lens, which focuses a designed pattern on photoresist (photosensitive coating of material) placed on a surface of a silicon wafer or film. Then, the exposed or the masked part of photoresist is removed leaving the desired pattern on silicon wafer or film depending on the characteristics of the used photoresist. To date, this technique has been widely used for fabricating nanoelectronics device. However, the dimension of nanopatterns which can be fabricated using this technique is limited by the wavelength of used light. Many improvements, including various technical improvements and modifications by using X-rays or extreme ultraviolet light, have been developed to reach resolution less than 20 nm. Therefore, those improvements make this technique rather expensive and increase its difficulties. Moreover, it seems that the resolution of nanopattern fabricated by lithography has reached near its limit.

Electron Beam Drawing or EB Drawing is the most potential candidate to replace current lithography technique. EB Drawing uses focused electron beam instead of light to bombard a thin resist layer on a surface of silicon or film followed by rinse process.

Depending on the characteristic of the resist, the bombarded parts will remain or be removed leaving nanopattern in accordance with CAD design. Hosaka and other researchers have reported the fabrication of nanodots with density more than 1 Tbit/in.^{2} or pitch of nanodots less than 25 nm. According to the theory, it is possible to fabricate nanodots with pitch less than 3-5 nm or density up to 83 Tbit/in.^{2}. However, many issues must be solved first including the used resist, proximity effect, focusing electron beam, and other technical problems. In Industry, EB Drawing is often used as application to fabricate a mask or template for lithography or stamper processes. However, EB Drawing is unlikely to be used in fabricating nanostructures at large and rapid scales due to very high cost, low throughput, and consuming high energy.

Ion Beam technique has similar principal with EB Drawing. Ion Beam technique uses ion beam instead of electron. One of techniques in this category, Ion Projection Lithography (IPL) has the complete absence of diffraction effects coupled with ability to tailor the depth of ion penetration to suit the resist thickness or the depth of modification. This characteristic gives advantage to pattern a large area in a single brief irradiation exposure without any wet processing step. While Focused Ion Beam (FIB) can produce master stamps in any material. 8 nm lines written into a multilayer sample of AlF3/GaAs have been succeeded fabricated using FIB. Although this technique has now breached the technologically difficult 100 nm barrier, and are now capable of fabricating structures at the nanoscale, the obstructions for fabricating high area and high density nanopattern with low cost production still remain. However, Ion Beam techniques are combined with nanoimprinting and pattern transfer for mass production.

In bottom-up methods, atoms or molecules are arranged to form higher dimension of nanostructures by mechanical/physical, chemical, or self-assembly processes. Bottom-up methods include nanofabrication using Scanning Probe Microscope, Chemical Vapour Deposition (CVD), and self-assembly process. Compare to others, self-assembly process has great advantages on high throughput and low cost. Among self-assembly techniques, however, block copolymer self-assembly technique is likely to be the most applicable in nanoelectronics device industry.

In general, nanofabrication techniques using SPM have been used for direct writing of nanostructures through material modification, deposition, and removal. The scanning tip of SPM acts as mechanical, thermal, and/or electric source to initiate and perform various physical and chemical for forming nanostructures. There are two major technologies within SPM which are used: scanning tunnelling microscopy (STM) and atomic force microscopy (AFM). Nanofabrication using STM has demonstrated manipulation of resist surface at atomic scale resolution to form sub-nano scale of pattern. Very low throughput, limited resist, and the requirements of high-vacuum and controlled environment have left this technique in laboratory. Using AFM, nanofabrication can be performed at normal ambient environment so more resist material can be used and the process is easier. However, the resolution of fabricated nanopatterns is limited by the size and by the shape of cantilever tip, and the tip writing speed (rate) is still need to be improved to increase its low throughput.

The product of Chemical Vapour Deposition (CVD) is almost deposit materials such as film in various forms with high purity and high performance. Some nanostructures fabrications are reported. However, the issue of order, alignment, and shape controls of nanostructures is still need to be resolved. Many process modifications and material selection in CVD are important to obtain applicable products. Therefore, we will not discuss this technique in detail.

The development of semiconductor technology has reached unprecedented level. The high-cost and limitation of nanofabrication using top-down methods yield the expectation of other methods increasing. Self-assembly technique offering low cost and ease process has attracted in recent few years. As an alternative technique, the self-assembly of block copolymer (BCP) offers a simple and low-cost process to form large-area periodic nanostructures, giving it great potential for application to fabricate patterned media that will be used for the next-generation magnetic recording. It have also been previously demonstrated that the self-assembled nanostructures have the potential to function as etching masks or templates in nano-patterned lithography. Aissou et. al. used BCP thin films as deposition and etching mask to fabricate silicon nanopilllar arrays with dimension and periodicities which are difficult using conventional technologies like optical, e-beam lithography and etching process. Therefore, this finding has increase possibility of BCP self-assembly technique to be utilized as an application for fabricating electronic devices.

Another significant achievement in the research to increase possibility of BCP self-assembly technique was reported recently. Bita et. al. used a topographical graphoepitaxy technique for controlling the self-assembly of BCP thin films that produces 2D periodic nanostructures with a precisely determined orientation and long-range order. The experiment showed that designated nano guide post formed by e-beam lithography, which is functionalized as surrogate nanodots of BCP self-assembly nanodots, effectively controlled nanodots in forming long range ordered nanodot pattern. In our research, Hosaka et. al. demonstrated the possibility of forming long-range ordering self-assembled nanodots array using mixing guide posts those are nanodots guide post and nano guide line fabricated by e-beam lithography. Recently, techniques to control orientation and periodicity of self-assembled nanostructures are well established by designated template, chemical modification, electric induction, modifying self-assembly medium, etc.

Two developed achievements in BCP self-assembled pattern-transfer and controlling periodicity and orientation of BCP self-assembled nanostructures have attracted many researchers and engineers in lithography field. This method offers an applicable new technique as strong candidate to replace current photolithography technique.

2. Self-assembly of block copolymer

A block copolymer consists of two or more polymeric blocks which are chemically or covalently bonded. In the melt condition, they are driven to segregate into various nanostructures by the repulsion of the immiscible blocks, almost in the case of a blend of immiscible homopolymer blocks. This segregation process, which gives result in microphase separation, is defined as self-assembly of block copolymer. In this far simpler case of block copolymer which consists of two homopolymer blocks, A and B, the nanostructure (phase) behaviour of BCP self-assembly process can be easily managed by simply varying parameters of the block copolymer, such as the total number of segments N, the volume fraction f of BCP component, the Flory-Huggins segmental interaction parameter χ, and the molecular architecture of the BCPs. Block copolymer consisting two homopolymer block is also called as diblock copolymer. The Flory–Huggins equation describes approximately how these parameters affect the free energy of a blend:

$$\frac{\Delta G_{mix}}{k_b T} = \frac{1}{N_A}\ln(f_A) + \frac{1}{N_B}\ln(f_B) + f_A f_B \chi \qquad (1)$$

The first two terms correspond to the configurational entropy of the system, and can be regulated via the polymerization chemistry to change the relative lengths of the chains and fractions of A versus B block of polymer. In the third term of (1), χ is associated with the non-ideal penalty of A–B monomer contacts and is a function of both the chemistry of the molecules and temperature. In general,

$$\chi = \frac{a}{T} + b \qquad (2)$$

where a and b are experimentally obtained constants for a given composition of a particular blend pair. Experimentally, χ can be controlled through temperature. Unlike macrophase separation in blends, the connectivity of the blocks in block copolymers prevents complete separation and instead the block copolymer chains organize to put the A and B portions on opposite sides of an interface. The equilibrium nanodomain structure must minimize unfavorable A–B contact without over-stretching the blocks. The strength of segregation of the two blocks is proportional to χN. A symmetric diblock copolymer is predicted to disorder (or pass through its order–disorder temperature (ODT)) when $\chi N < 10$. Below the ODT and when the volume fraction of block A (f_A) is quite small, it forms spheres in a body-centered cubic (BCC) lattice surrounded by a matrix of B. As f_A is increased towards 0.5, the minority nanodomains will form first cylinders in a hexagonal lattice, then a bicontinuous double gyroid structure, and finally lamellae. This phenomenon is described in Fig. 1

Mean-field phase diagram in Fig. 1 shows the theory of BCP self-assembled nanostructures. The X-axis corresponds to the volume fraction of block copolymer f_A, and the Y-axis corresponds to the product of the Flory-Huggins parameter χ and the total number of segments N. As shown in Fig. 1, nanostructures such as spherical, cylindrical, or lamellar structures can be obtained by changing the volume fraction f of BCP components. Other structures such as gyroid and close-packed sphere (CPS) structures are also observed on a small range of certain BCP composition. Fig. 1 also shows that the product of χN in the Y-axis is relatively equivalent to the self-assembled nanostructure size due to the total number of BCP segments N equal to the size of BCP. This means that to obtain smaller nanostructure by using the same BCP, it is necessary to adopt the same BCP with lower total number of BCP segments N. In experiment, we chose BCP with low molecular weight to obtain smaller nanodots.

In this chapter, fabrication of nanodots using block copolymer technique will be discussed. The discussion is based on experiment of block copolymer self-assembly technique using Polystyrene-Poly(dimethylsiloxane) (PS-PDMS) block copolymer. We demonstrated the nanodot pattern fabrication using polystyrene-poly(dimethyl siloxane) (PS-PDMS) BCPs with three different molecular weights of 30,000-7,500, 13,500-4,000, and 11,700-2,900. We chose PS-PDMS because of its high etch selectivity and its high Flory-Huggins parameter χ. PS-PDMS block copolymers have been demonstrated to have a high etch selectivity between the two kinds of blocks because organic PS block can be easily etched while Si-containing organometallic PDMS block as nanodots shows high endurance to etching. PDMS which contains Si will form a material with property like silica when etched by oxygen (O_2) gas. This result implies on high etch selectivity between PDMS and PS domain. Its high etch selectivity will give advantage in subtractive pattern transfer while its Flory-Huggins parameter χ implies its advantage in obtaining very fine nanodot pattern with pitch size at sub-10 nanometers. In practice, PS-PDMS has economic advantages since the annealing temperature and annealing time, which is necessary to promote self-assembly process, are relatively less compare to BCP which is used in conventional research such as PS-PMMA. It was also reported that vapour solvent-annealing treatment was demonstrated to promote self-assembly process of PS-PDMS.

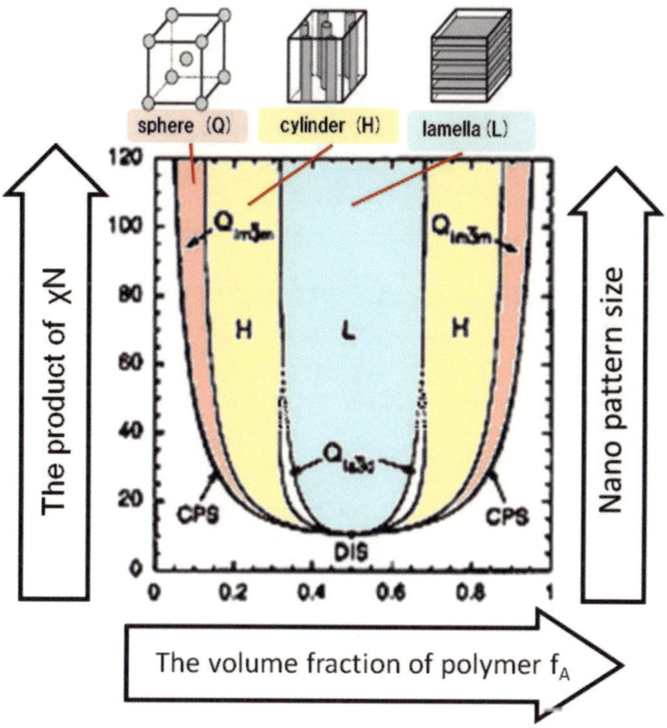

Fig. 1. Mean-field phase diagram for BCP melts shows the theory of BCP self-assembled nanostructures. The images in above diagram describe schematics of thermodynamically stable block copolymer phases.

3. Experimental method

In this work, we adopted spherical morphology PS-PDMS BCPs with molecular weights of 30,000-7,500, 13,500-4,000, and 11,700-2,900 and minority block volume fraction f_{PDMS} of 20%, 24%, and 20.9%, respectively. Fig. 2(a) schematically shows the experimental process that is used to form self-assembled nanodots. We dissolved PS-PDMS in a certain solvent to obtain PS-PDMS solutions with a weight concentration of 2%. Then, PS-PDMS solutions were spin-coated onto silicon substrates of 1 cm^2 and annealed at 170°C in a N_2 atmosphere or vacuum for 11 hours.

Fig. 2(b) illustrates the process of self-assembly using PS-PDMS. The spherical PDMS nanodots formed by microphase separation are shown here. Fig. 2(b-1) shows a spin-coated PS-PDMS film sample. After annealing, a very thin PDMS layer preferentially segregates at the air/polymer interface because of the low surface energy of PDMS. This produces the structure schematically shown in Fig. 2(b-2). We conducted reactive ion etching (RIE) using CF_4 gas to remove the very thin top PDMS layer. Finally, we conducted RIE using O_2 gas in order to remove the PS domain and to form the nanodot pattern. This O_2 plasma etching also causes the Si-containing PDMS domain partly oxidized, leaving a material with properties similar to silica that is robust for subtractive pattern transfer. RIE was conducted for all samples in the same condition. The etched nanodots are schematically shown in Fig. 2(b-3).

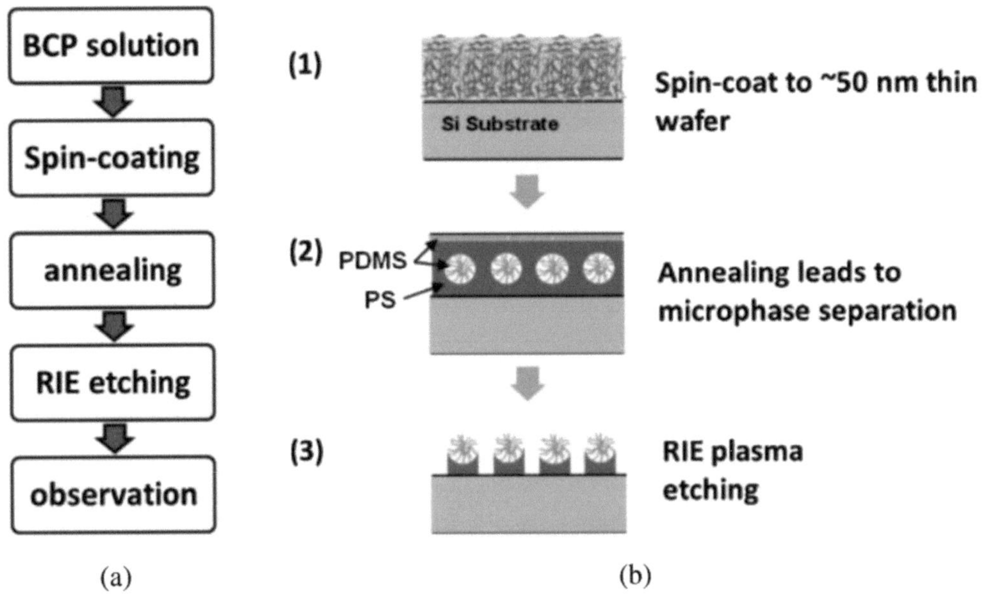

Fig. 2. Self-assembly process of BCP: (a) experimental process flow and (b) self-assembly process of PS–PDMS BCP.

We used a high-resolution scanning electron microscope (SEM; JEOL JSM6500F) to observe the PS-PDMS films after etching process. We used atomic force microscopy (AFM) to measure the thickness of the PS-PDMS film. The PS-PDMS films were partly removed, and the difference in height was measured. This measurement was conducted before annealing was applied on the sample.

4. Result and discussion

In our first experiment, we chose PS-PDMS with molecular weight of 30,000-7,500 and minority block volume fraction f_{PDMS} of 20%. In experiment, this PS-PDMS formed nanodots pattern with optimal size approximately 33 nm in pitch and 23 nm in diameter. Then, we selected PS-PDMS with lower molecular weights of 13,500-4,000 and 11,700-2,900 to obtain smaller nanodots.

4.1 Formation of nanodots formed by PS-PDMS 30,000-7,500

Firstly, we investigated the dependence of self-assembly process on concentrations. Figs. 3(a), (b), and (c) show SEM images of fabricated nanodots by PS-PDMS self-assembly with different concentrations (wt/wt) of 1%, 2%, and 3%, respectively. According to PS-PDMS self-assembly process shown in the Fig. 2(b), nanodot-shaped patterns and the surrounding regions in Fig. 3 are believed to correspond to unetched-PDMS domain and bared Si substrate after PS matrix was removed by O_2 RIE. It can be seen from the SEM image of Fig. 3 (b) that the nanodots look darker than the surroundings, compared with Figs. 3(a) and 3(c). The main reason for this should be the so-called edge effect that more secondary electrons are produced at edges and thus the regions at edges look brighter than other regions. The relatively narrow gap between nanodots shown in Fig. 3(b) enhances the edge effect, making the gaps look brighter than nanodots. On the contrary, the gaps are relatively wide for formed nanodots shown in Figs. 3(a) and 3(c) so the gap regions reasonably look dark although the dark nanodots are surrounded by brighter edges.

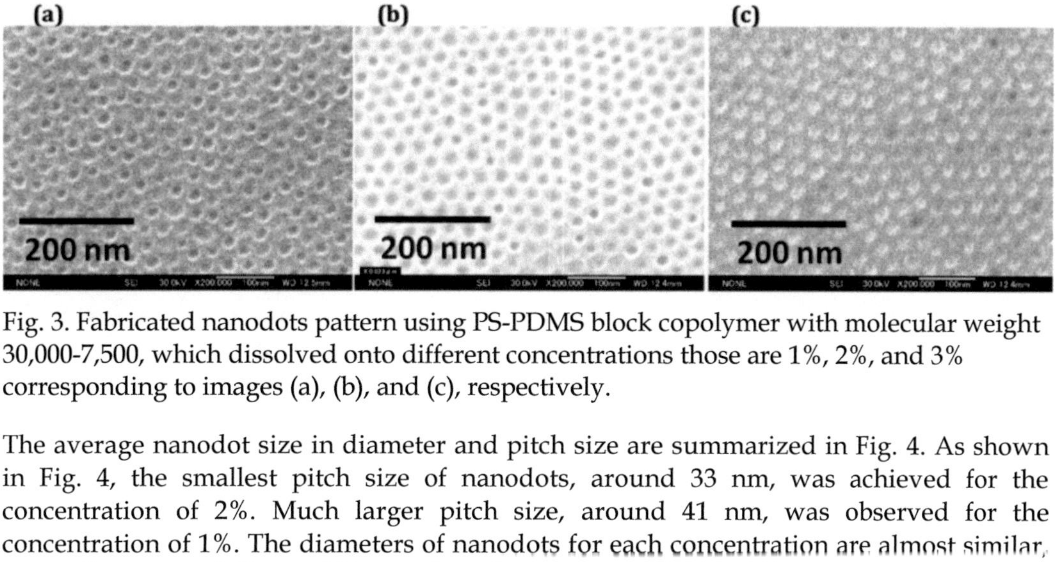

Fig. 3. Fabricated nanodots pattern using PS-PDMS block copolymer with molecular weight 30,000-7,500, which dissolved onto different concentrations those are 1%, 2%, and 3% corresponding to images (a), (b), and (c), respectively.

The average nanodot size in diameter and pitch size are summarized in Fig. 4. As shown in Fig. 4, the smallest pitch size of nanodots, around 33 nm, was achieved for the concentration of 2%. Much larger pitch size, around 41 nm, was observed for the concentration of 1%. The diameters of nanodots for each concentration are almost similar, as small as around 23 nm. This result makes us conclude that 2% in concentration is the most suitable concentration to optimize the pitch size of nanodots into the smallest, which indicates high density of nanodots.

Fig. 5 shows the SEM images of fabricated nano-patterns as a function of dropped volume. Figs. 5 (a), (b), and (c) show the SEM images of these patterns formed by dropping different volumes of 10, 20, and 40 μL onto Si substrate with a size of 1 cm^2, respectively. Only 2% PS-PDMS solution was used here. The enhanced edge effect can be also observed in Fig. 4(c) when the gap between nanodots is relatively narrow. When the dropped volumes of PS-PDMS solution are 20 μL or 40 μL, the morphologies of PS-PDMS self-assembly pattern are spherical structure. The morphology of PS-PDMS self-assembled pattern becomes fingerprint-like cylindrical structure when the dropped volume of PS-PDMS solution was reduced to 10 μL. The width of the cylindrical structure is around 16 nm, similar to our experimental result of PS-PDMS self-assembly with a molecular weight of 13,500-4,000.

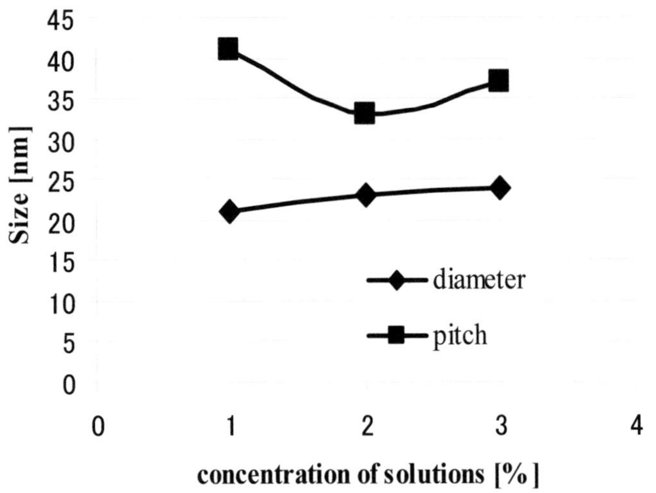

Fig. 4. Relation between nanodot size and concentration.

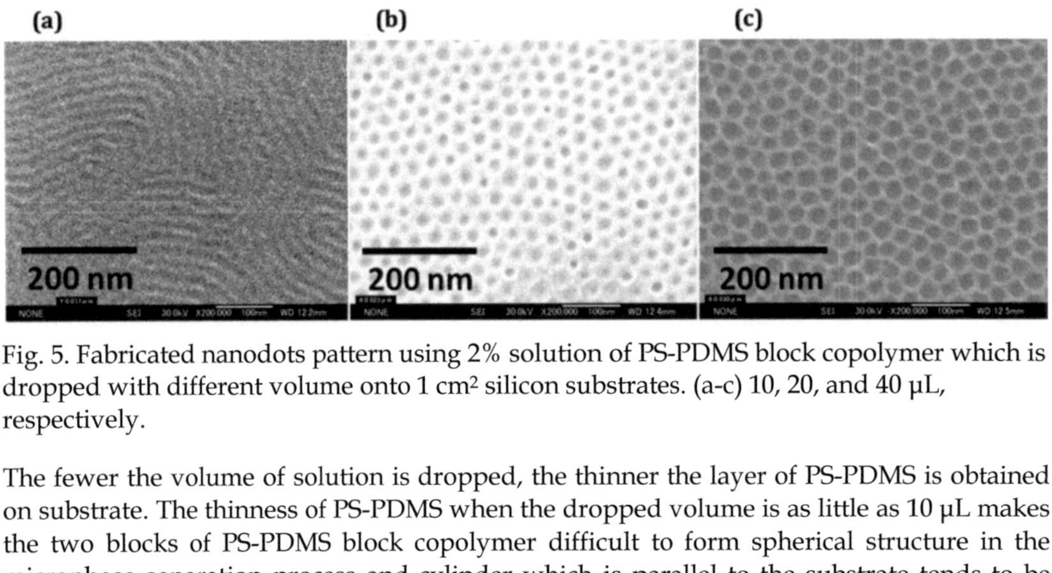

Fig. 5. Fabricated nanodots pattern using 2% solution of PS-PDMS block copolymer which is dropped with different volume onto 1 cm^2 silicon substrates. (a-c) 10, 20, and 40 μL, respectively.

The fewer the volume of solution is dropped, the thinner the layer of PS-PDMS is obtained on substrate. The thinness of PS-PDMS when the dropped volume is as little as 10 μL makes the two blocks of PS-PDMS block copolymer difficult to form spherical structure in the microphase separation process and cylinder which is parallel to the substrate tends to be formed. We believe that this phenomenon should be related to the lowest system energy. Forming nanodots for the little dropped volume must require higher energy than forming cylindrical structures. We summarized this tendency from Fig. 5 into graph shown in Fig. 6.

Fig. 6 also shows that there is correlation between the dropped volume of PS-PDMS block copolymer solutions and the size of formed nanodots. Since the dropped volume of PS-PDMS block copolymer solutions is corresponding to the thickness of PS-PDMS block copolymer layer, we believe that the size of nanodots can be controlled simply by dropping the most proper volume of block copolymer solutions. The thicker the PS-PDMS block copolymer layer is, the bigger the size of the formed nanodots is, although the size of nanodots is limited by the size of nanodot pitch.

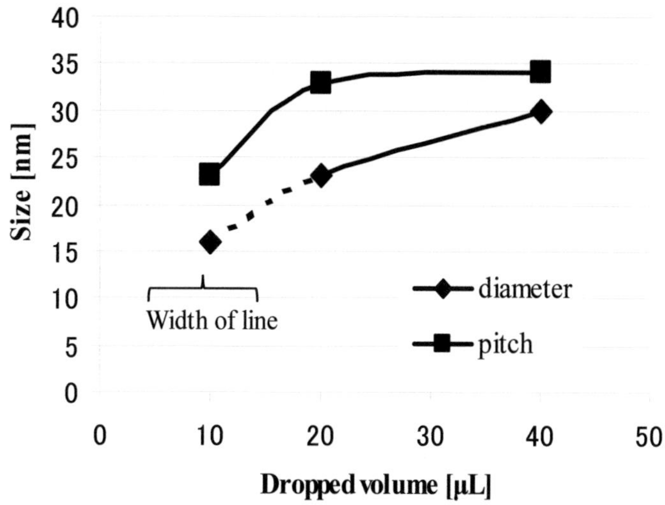

Fig. 6. Relations between the dropped volume of PS-PDMS solution and the size of nanodots. The dashed line indicates that the pattern morphology alters into random line.

4.2 Formation of smaller nanodots

In our first experiment, we used PS-PDMS which has molecular weight of 30,000-7,500 and minority volume fraction f_{PDMS} of 20%. By optimizing experimental parameters, we succeeded in forming nanodot pattern with the pitch of nanodots as small as approximately 33 nm and the diameter of nanodots as small as 23 nm. In order to obtain smaller nanodots pattern, it is necessary to use BCP with less molecular weight and to keep the value of BCP's minority volume fraction according to the mean-field phase diagram for block copolymer melts. Therefore, we adopted two PS-PDMS BCPs which have molecular weights of 13,500-4,000 and 11,700-2,900 and minority block volume fractions f_{PDMS} of 24% and 20.9%, respectively.

PS-PDMS has high Flory-Huggins parameter χ of 0.26. In the strong segregation of BCP such as PS-PDMS with a molecular weight of 30,000-7,500 and with χN value of 101.2 (the molecular weights of styrene and dimethyl-siloxane blocks are 104.15 and 74.15, respectively), the pitch size of self-assembled nanodots is given by a formula below.

$$\text{Pitch} = aN^{\frac{2}{3}}\chi^{\frac{1}{6}} \tag{3}$$

where a is segment length. To predict the pitch size of PS-PDMS with molecular weights of 13,500-4,000 and 11,700-2,900, we used formula (3) by assuming that the estimation was done with the following two conditions being met. The first condition was that the segment lengths of the two adopted PS-PDMS are near the segment length of PS-PDMS with molecular weight of 30,000-7,500. The second was that both should promote microphase separation in the strong segregation. The prediction for the pitches of nanodots is shown in Fig. 7. It was predicted that PS-PDMS BCPs with molecular weights of 13,500-4,000 and 11,700-2,900 would form nanodots with pitch sizes as small as 24.3 nm and 17.5 nm, respectively.

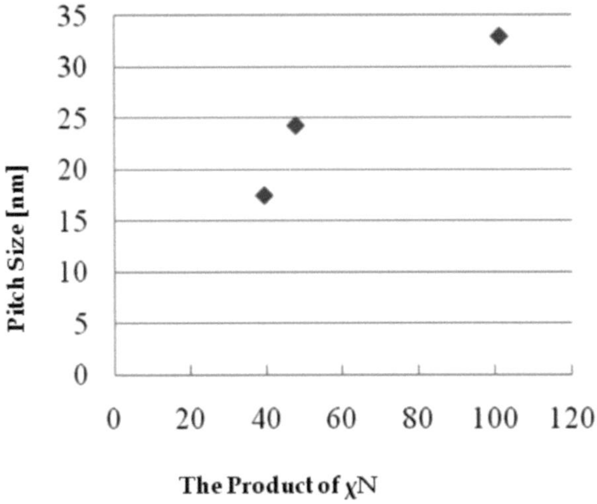

Fig. 7. The prediction of nanodots pitches of PS–PDMS BCPs with molecular weights of 30,000–7,500, 13,500–4,000, and 11,700–2,900.

4.3 Formation of nanodots formed by PS-PDMS 13,500-4,000

Fig. 8 shows SEM images [Fig. 8(a)] of self-assembled nanopatterns using PS-PDMS BCP with molecular weight of 13,500-4,000 and the corresponding schematic cross-section images

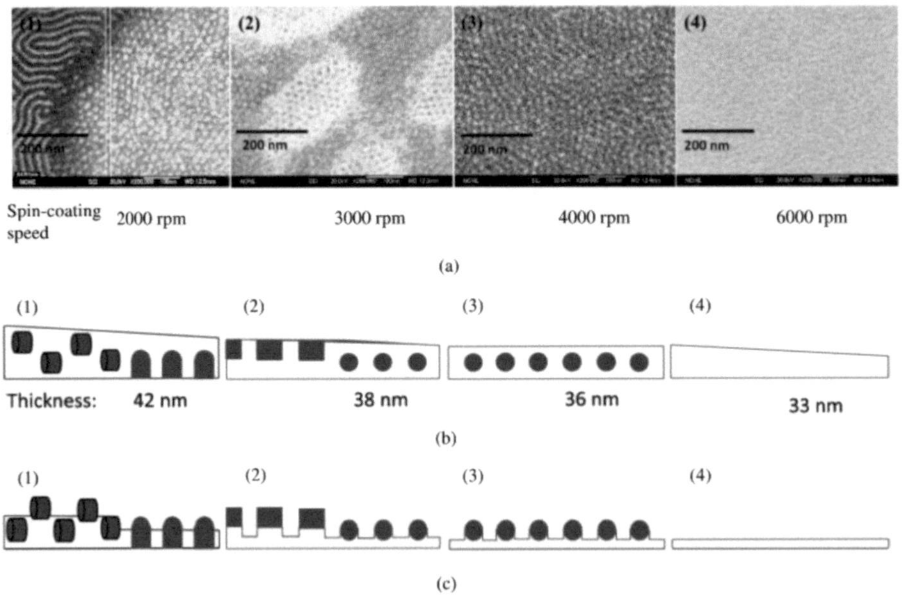

Fig. 8. (a) SEM images of self-assembled nanopatterns using PS–PDMS BCP with the molecular weight of 13,500–4,000 and the corresponding cross- section images of their schematic height profile. Images (1)–(4) show SEM images of nanopatterns when the spin-coating speed was changed to 2000, 3000, 4000, and 6000 rpm, respectively. (b, c) Schematic cross-section images of the correlating film height profile: (b) before etching and (c) after etching.

of the film height profile before [Fig. 8(b)] and after [Fig. 8(c)] etching process. Fig. 8(a) shows SEM images of self-assembled nanopatterns when the spin-coating speed changed to (1) 2000, (2) 3000, (3) 4000, and (4) 6000 rpm. All these images were taken on samples after O_2 plasma etching had been conducted on samples. The thickness of the PS-PDMS films, which formed before etching process, decreased with the increase of the spin-coating speed, as shown in Fig. 8(b). The thickness of the PS-PDMS films varies with spin-coating speed, and it is possible to obtain the optimum thickness for self-assembly by changing the spin-coating speed. Fig. 8(c) shows the schematic cross-section images of the film height profile after etching process, the numbers correspond to those of Figs. 8(a) and 8(b).

When the spin-coating speed was 2000 rpm, the thickness of the PS-PDMS film was 42 nm and the self-assembled nanostructure of the PDMS cylinders that was perpendicular and parallel to the substrate surface was obtained, as shown in Fig. 8(a-1). The direction of the parallel cylinders was partially orientated. According to our analysis, the width and pitch of the cylinders were 12 and 22 nm, respectively. However, the PDMS cylinders perpendicular to the substrate surface looked like nanodots when observed by SEM. The diameter and pitch of these nanodots were 12 and 22 nm, respectively.

When the spin-coating speed was 3000 rpm, the thickness of the PS-PDMS film was 38 nm, and self-assembled nanodots and nanoholes as a result of the semi-perforated lamella nanostructure were obtained. The brighter dots, which are surrounded by the darker region, are nanodots, and the darker dots in the brighter region are nanoholes. According to our analysis, the diameter and pitch of both nanodots and nanoholes were 12 and 22 nm, respectively. This result is shown in Fig. 8(a-2).

When the spin-coating speed was 4000 rpm, the thickness of the PS-PDMS film was 36 nm, and self-assembled nanodots were formed. The diameter and pitch of the nanodots were roughly 12 and 22 nm, respectively. This result is shown in Fig. 8(a-3).

When the spin-coating speed was increased to 6000 rpm, the thickness of the film decreased to 33 nm and no nanodots formed. This result is shown in Fig. 8(a-4).

The uniformity of the nanodots on a large area was also investigated. SEM images were randomly taken on a 1 cm^2 silicon substrate after etching process. The nanodot arrays formed uniformly when a 36-mm thick PS-PDMS film was formed with a spin-coating speed of 4000 rpm.

As shown in Fig. 8(a), the self-assembled nanostructures obtained when using PS-PDMS with molecular weight of 13,500-4,000 are not morphologically stable. We believe that this phenomenon occurred because the value of the PDMS fraction of this PS-PDMS is near the value of that which assembled into the cylinder nanostructure. This hypothesis also applies to the self-assembled nanostructure, which consists of cylinders perpendicular and parallel to the substrate surface, and was formed using PS-PDMS with molecular weight of 13,500-4,000 as shown in Fig. 8(a-1).

We describe the formed nanostructures when using PS-PDMS with molecular weight of 13,500-4,000, which change systematically as a function of the gradually changing film thickness in Fig. 8(b) (before etching) and in Fig. 8(c) (after etching), respectively. The darker regions indicate the PDMS block domains. When the film thickness was 42 nm, the hybrid nanostructure of the PDMS cylinders, which are perpendicular and parallel to the substrate surface, were formed. This corresponds to the experimental result in Fig. 8(a-1). When a thinner film thickness of 38 nm was spin-coated, nanodots and semi-perforated lamella were formed. At the semi-perforated lamella, the minority block of PDMS fills the lamella domain at the surface of film, which is perforated by the majority block of PS filling it below. The short duration of CF_4 etching, however, was not enough to remove the minority block of PDMS at the surface. The remaining PDMS domains acted as a mask when the RIE etching of O_2 was conducted to remove the PS domain, which caused nanohole structures to be formed. This corresponds to the experimental result in Fig. 8(a-2), where the nanodots are shown as brighter dots and the nanoholes are shown as darker dots. Those results are consistent with a report by A. Knoll et al. that film thickness affects the formed nanostructures, which are determined by an interplay between surface field and confinement effect. The obtained nanostructures of cylinder-forming block copolymer were confirmed by experimental results and simulation using dynamic density functional theory.

When the optimum thickness of the PS-PDMS film was achieved, self-assembled nanodots were able to form on a large area, as shown in Fig 8(a-3). In the present experiment, the optimum thickness is 36 nm to form one nanostructure of nanodots, as described in Fig. 8(b-3). When the thickness of the film is less than the optimum thickness, no nanostructure is formed, as shown in Fig. 8(b-4). This corresponds to the experimental result in Fig. 8(a-4) when the film thickness was 33 nm. We believe that the confinement effect and the block length factor of PS-PDMS prevented microphase separation, and caused either a disordered phase or a lamellar wetting layer, which is formed at this very small thickness.

4.4 Formation of nanodots formed by PS-PDMS 11,700-2,900

Fig. 9(a) shows SEM images of self-assembled nanopattern using PS-PDMS with the molecular weight of 11,700-2,900 and the corresponding schematic cross-section images of the film height profile before [Fig. 9(b)] and after [Fig. 9(c)] etching process. Fig. 9(a) shows SEM images of nanopatterns when the spin-coating speed was changed to (1) 4000, (2) 6000, and (3) 8000 rpm. All these images were also taken on samples after O_2 plasma etching had been conducted on samples. According to our analysis, self-assembled nanodot patterns with diameter and pitch as small as 10 and 20 nm, respectively, were obtained. Similar to the PS-PDMS of 13,500-4,000, the thickness of the PS-PDMS film, which formed before etching process, decreased with the increase of the spin-coating speed. The thickness of the PS-PDMS film significantly decreased when the spin-coating speed increased from 4000 to 6000 rpm. Fig. 9(c) shows the schematic cross-section images of the film height profile after etching process, and the numbers correspond to those of Figs. 9(a) and 9(b).

When the spin-coating speed was 4000 rpm, self-assembled nanodots with diameter and pitch as small as 10 and 20 nm, respectively, were obtained. This result is shown in Fig. 9(a-1). However, the thickness of the PS-PDMS film was 51 nm, which was more than two times the pitch of the nanodots. We believe that two layers of self-assembled nanodots were formed on the surface of the substrate, as described in Fig. 9(b-1). Therefore, the SEM image in Fig. 9(a-1) only shows the first layer of the self-assembled nanodots near the surface of PS-PDMS film as shown in Fig. 9(c-1) since the same condition of RIE was applied on all samples.

When the spin-coating speed was 6000 rpm, the thickness of the PS-PDMS film was 33 nm and self-assembled nanodots were formed. According to our analysis, the diameter and pitch of the nanodots were 10 and 20 nm, respectively. This result is shown in Fig. 9(a-2).

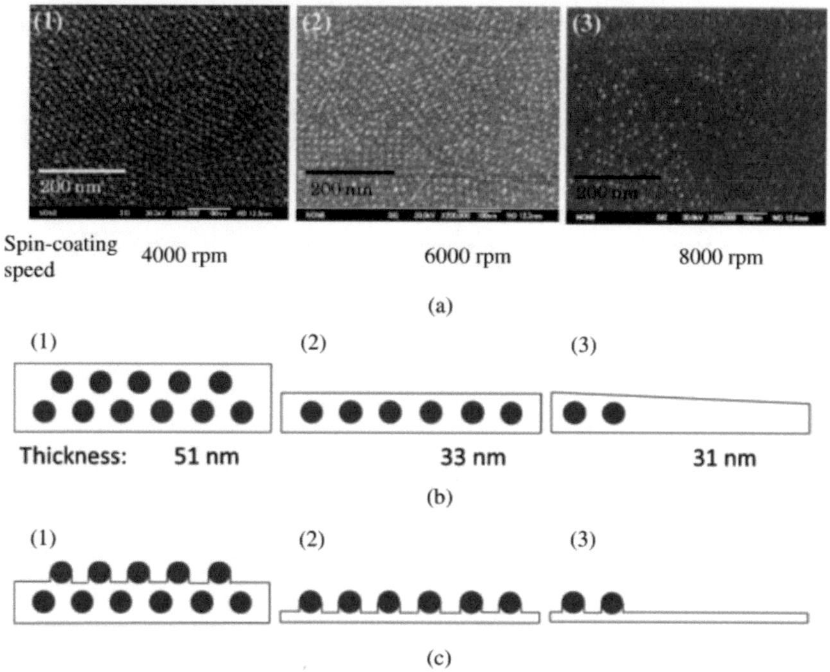

Fig. 9. (a) SEM images of self-assembled nanopatterns using PS–PDMS with the molecular weight of 11,700–2,900. Images (1)–(3) show SEM images of nanopatterns when the spin-coating speed was changed to 4000, 6000, and 8000 rpm, respectively. (b, c) Schematic cross-section images of the correlating film height profile: (b) before etching and (c) after etching.

When the spin-coating speed was 8000 rpm, the thickness of the PS-PDMS film was 31 nm and self-assembled nanodots were arbitrarily formed on the surface of the substrate. The diameter of the nanodots was 10 nm. This result is shown in Fig. 9(a-3).

The self-assembled nanodots were well-formed using PS-PDMS with molecular weight of 11,700-2,900, as shown in Figs. 9(a-1) and 9(a-2). However, nanodots in Fig. 9(a-1) were self-assembled into two layers since the thickness of the PDMS film was double the pitch of the nanodots. Hence, this nanodot pattern would be difficult to be utilized on the pattern transfer. As shown in Fig. 9(a-3), the self-assembled nanodots were partly formed on the surface of substrate. The reason of this is because the thickness of PDMS film is too thin.

According to the results, the pitch of nanodots obtained experimentally is different from our prediction, as shown in Fig. 10. In our experiment, the pitch is directly proportional to the product of χN, as shown by the dashed line in Fig. 10. This difference is probably caused by two factors. The first is inappropriate value of the segment length and the second is that the microphase-separation condition of used PS-PDMS is not in strong segregation. This graph, however, could be used as a guide in the selection of the molecular weight of PS-PDMS BCP to form the smaller self-assembled nanodots.

According to our experiment result that PS-PDMS with molecular weight of 11,700-2,900 is more stable in forming nanodots than PS-PDMS with molecular weight of 13,500-4,000 due to the value of its PDMS volume fraction, and we describe this phenomenon in mean-field phase diagram in Fig. 11. We believe that the dotted-line separating between the cylinder-forming PS-PDMS and the sphere-forming PS-PDMS could be drawn near PS-PDMS with molecular weight of 13,500-4,000 because the characteristic of this PS-PDMS is unstable in forming sphere structure and tends to form cylinder. The dashed line separates the order-disordered phase of PS-PDMS. Therefore, the region between the dotted line and the dashed line describes the spherical morphology of PS-PDMS BCP. Fig. 11, however, is important in the selection of the PS-PDMS molecular weight and the minority volume block fraction of PDMS in order to form the desired nanopattern. Therefore, it is necessary to adopt PS-PDMS with minority volume block fraction of PDMS that is smaller than 24% to obtain nanodots pattern according to our experimental result.

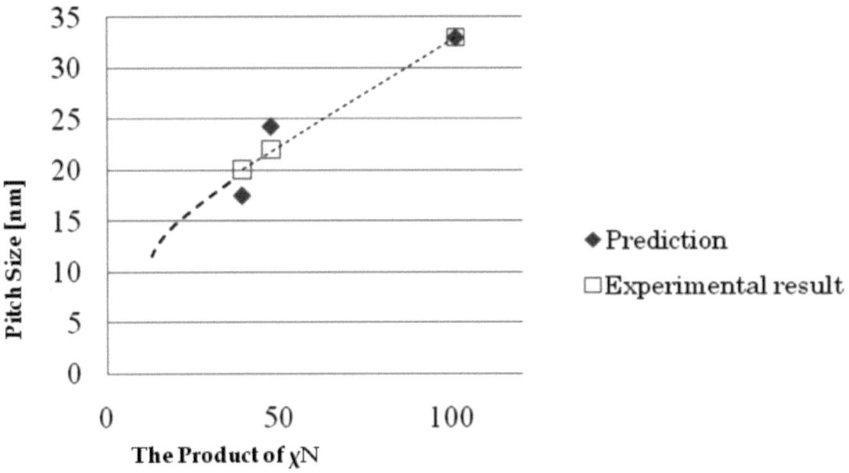

Fig. 10. A graph of the pitch sizes of nanodots according to prediction and experimental results.

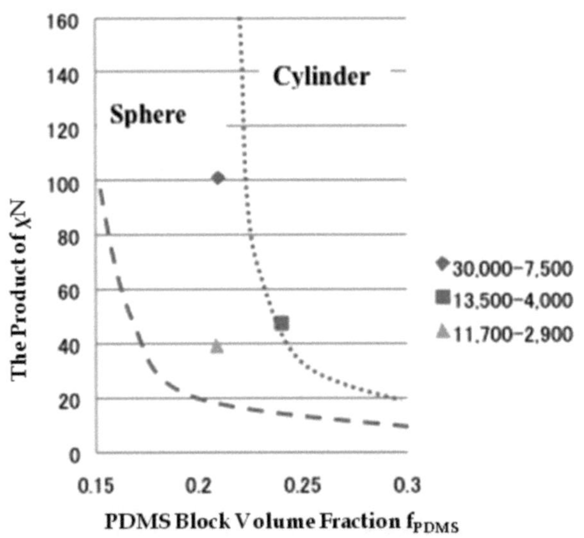

Fig. 11. The position of our adopted PS-PDMS in mean-field phase diagram for BCP melts.

We have succeeded in fabricating self-assembled nanodots with a diameter as small as 10 nm and a pitch size as small as 20 nm using PS-PDMS with molecular weight of 11,700-2,900. Those nanodots were able to form on a large area, which was confirmed by randomly imaging the surface of a 1 cm^2 sample using SEM. However, PS-PDMS with molecular weight of 11,700-2,900 is more stable in forming nanodots than PS-PDMS with molecular weight of 13,500-4,000 due to the value of its PDMS fraction. The nanodot size promises for fabricating 1.86 Tbit/$in.^2$ storage device using patterned media method by ease process and low cost fabrication. It is advantageous that the gap between nanodots is the same as the nanodot size, as this will reduce the effect of instability from thermal fluctuation.

5. Conclusion

We have succeeded in fabricating self-assembled nanodots using PS-PDMS with molecular weights of 30,000-7,500, 13,500-4,000, and 11,700-2,900. By optimizing the concentration and the dropped volume of PS-PDMS solution, nanodots with a diameter as small as 23 nm and a pitch size as small as 33 nm were formed using PS-PDMS with molecular weights of 30,000-7,500. These indicate that the concentration and the dropped volume of PS-PDMS solution play role on the formed nanopattern size.

Nanodots with a diameter as small as 12 nm and a pitch size as small as 22 nm were formed using PS-PDMS with molecular weights of 13,500-4,000, and nanodots with a diameter as small as 10 nm and a pitch size as small as 20 nm using PS-PDMS with molecular weight of 11,700-2,900. The latter nanodot size is promising for fabrication 1.86 Tbit/$in.^2$ storage device because of its high possibility to be used as mask for subtractive transfer process. In experiment, we found that an optimum thickness of the PS-PDMS film is required to obtain self-assembled nanadots on large areas especially when the position of the BCP is not in stable area in mean-field phase diagram for BCP melts. The optimum thickness could be obtained simply by adjusting the spin-coating speed during formation of the thin block copolymer film. We have predicted the limitations in the selection of PS-PDMS BCP to be used in forming smaller sizes of self-assembled nanodots of this BCP. In order to increase the possibility of applying this technique to nano-electronic devices, we plan on further reducing the nanodot size and controlling the orientation of nanodots in the future.

6. Acknowledgment

This work was funded by the New Energy and Industrial Technology Development Organization (NEDO) under the development of nanobit technology for the ultrahigh density magnetic recording (Green IT) project. We gratefully thank Prof. K. Itoh from Graduate School of Engineering, Gunma University, for the use of RIE tool.

7. References

Huda, M.; Yin, Y. & Hosaka, S. (2010) Self-assembled nanodot fabrication by using diblock Copolymer. *Key Eng. Mater. AMDE,* Vol. 459, pp. 120-123, ISSN 1662-9795

Huda, M.; Akahane, T.; Tamura, T.; Yin, Y. & Hosaka, S. (2011) Fabrication of 10-nm-order block copolymer self-assembled nanodots for high-density magnetic recording. *Jpn. J. Appl. Phys.,* Vol. 50, pp. 06GG06-1 - 06GG06-5, ISSN 1347-4065

Hosaka, S.; Zulfakri, B. M.; Shirai, M.; Sano, H.; Yin, Y.; Miyachi, A. & Sone, H. (2008) Extremely small proximity effect in 30 keV electron beam drawing with thin calixarene resist for 20×20 nm^2 pitch dot arrays. *Appl. Phys. Express,* Vol. 1, pp. 027003-027005, ISSN 1882-0786

Hieda, H.; Yanagita, Y.; Kikitsu, A.; Maeda, T. & Naito K. (2006) Fabrication of FePt patterned media with diblock copolymer templates. *J. Photopolym. Sci. Technol.* Vol. 19, No. 3, pp. 425-430, print: ISSN 0914-9244, online: ISSN 1349-6336

Matsen, M. W. & Schick, M. (1994) Stable and unstable phases of a diblock copolymer melt. *Phys. Rev. Lett.*, Vol. 72 pp. 2660–2663, print: ISSN 0031-9007, online: ISSN 1079-7114

Ross, C. A. et al. (2008) Si-containing block copolymers for self-assembled nanolithography. *J. Vac. Sci. Technol. B.* Vol. 26, pp. 2489-2494, print: ISSN 1071-1023, online: ISSN 1520-8567

Bita, I.; Yang, J. K. W.; Jung, Y. S.; Ross, C. A.; Thomas, E. L.; & Berggren, K. K. (2008) Graphoepitaxy of Self-Assembled Block Copolymers on Two-Dimensional Periodic Patterned Templates. *Science*, Vol. 321, pp. 939-943, print: ISSN 0036-8075, online: ISSN 1095-9203

Knoll, A.; Horvat A.; Lyakhova, K. S.; Krausch, G.; Sevink, G. J. A.; Zvelindovsky, A. V. and Magerle R. (2002) Phase Behavior in Thin Films of Cylinder-Forming Block Copolymers. *Phys. Rev. Lett.*, Vol. 89, pp. 035501-035504, print: ISSN 0031-9007, online: ISSN 1079-7114

Bates, F. S. & Fredrickson, G. H. (1990) Block Copolymer Thermodynamics: Theory and Experiment. *Annu. Rev. Phys. Chem.,* Vol. 41, pp. 525-557, ISSN 0066-426X

Aissou, K.; Kogelschatz, M.; Baron, T. and Gentile, P. (2007) Self-assembled block polymer templates as high resolution lithographic masks. *Surf. Sci.* Vol. 601, pp. 2611-2614, ISSN 00396028

6

Formation of Ultrahigh Density Quantum Dots Epitaxially Grown on Si Substrates Using Ultrathin SiO_2 Film Technique

Yoshiaki Nakamura[1] and Masakazu Ichikawa[2]
[1]*Graduate School of Engineering Science, Osaka University*
[2]*Department of Applied Physics, The University of Tokyo*
Japan

1. Introduction

Development of Si-based light emitter has been eagerly anticipated in Si photonics. However, its realization is difficult because group IV semiconductors such as Si and Ge are indirect-transition semiconductors. Si or Ge quantum dots (QDs) on Si substrates have drawn much attention as Si-based light emitting materials because their optical transition probability can be enhanced by their quantum confinement effect. There are some kinds of QDs fabricated by various methods: Stranski Krastanov (SK) QDs (Eaglesham & Cerullo, 1990; Schmidt & Eberl, 2000), Ge nanoparticles in SiO_2 matrix (Maeda, 1995), Si QDs by anodic oxidation (porous Si) (Wolkin et al., 1999; Cullis & Canham, 1991), and so on. In terms of the crystal orientation control, SK QDs have intensively attracted much interest. In general, the density of SK QDs is approximately $10^{10\text{-}11}$ cm^{-2} and the size is about 50-100 nm. In order to get strong light emission and quantum confinement effect, the higher density and smaller size are required.

We have developed a formation method of QDs using the ultrathin SiO_2 films which we call as ultrathin SiO_2 film technique (Shklyaev et al., 2000; Shklyaev & Ichikawa, 2001; Nakamura et al., 2004). In this method, Si or Ge QDs with ultrahigh density ($>10^{12}$ cm^{-2}) and small size (<5 nm) were epitaxially grown on Si substrates, where lattice mismatch strain was not used for the formation of QDs unlike SK QDs. Furthermore, QDs were elastically strain-relaxed without misfit dislocations (Nakamura et al., 2010, Nakamura et al, 2011b). These high crystal quality ultrasmall QDs on Si can be expected as Si-based light emitting materials.

In order to use QDs on Si substrates as light emitter in Si photonics, the light wavelength corresponding to energy bandgap E_g has to be consistent with that for optical fiber communications: namely 1.3 μm (~0.95eV) or 1.5 μm (~0.8 eV) wavelength bands. Ge bulk has E_g of 0.67 eV at room temperature, but E_g of Ge QDs increases up to 1-1.5 eV at QD diameter of around 5 nm due to quantum confinement effect (Niquet et a.l, 2000; Nakamura et al, 2005). On the other hand, $Ge_{1-x}Sn_x$ alloy films have been reported to be direct-transition semiconductors (Jenkins & Dow, 1987; He & Atwater, 1997) at larger Sn content x

(x>~0.12) indicating the possibility of high light emitting efficiency. In terms of energy bandgap, however, it is much smaller (0.3–0.5 eV) at x of ~0.12 (de Guevara et al., 2004) than that for optical fiber communications. We noticed a possibility that the energy bandgap of the direct-transition semiconductor $Ge_{1-x}Sn_x$ QDs could increase up to ~0.8 eV using their quantum confinement effect.

The growth of $Ge_{1-x}Sn_x$ alloys is complicated by a limited mutual solid solubility of Ge and Sn (~1%), and a tendency for Sn surface segregation. Moreover, the epitaxial growth of high crystal quality $Ge_{1-x}Sn_x$ alloys on Si is difficult because of the large lattice mismatch between $Ge_{1-x}Sn_x$ alloys (x >~0.12) and Si. Development of formation technique of $Ge_{1-x}Sn_x$ QDs was required. In this chapter, we try to solve the above problems in $Ge_{1-x}Sn_x$ QD formation by modifying our ultrathin SiO_2 film technique and develop the epitaxial growth technique of ultrahigh density (>10^{12} cm^{-2}) $Ge_{1-x}Sn_x$ QDs at high Sn content (x>~0.1), where QDs have almost no misfit dislocation and ultrasmall size causing quantum confinement effect. We present the results of formation and physical properties of QDs epitaxially grown on Si

2. The formation technique of Si and Ge QDs using ultrathin SiO_2 films

We have developed the formation technique of QDs using ultrathin SiO_2 films (Shklyaev et al., 2000; Shklyaev & Ichikawa, 2001; Nakamura et al., 2004). Si or Ge deposition on the ultrathin SiO_2 films brings epitaxial growth of ultrasmall QDs with ultrahigh density (>10^{12} cm^{-2}). This growth mode is totally different from the conventional SK and Volmer Weber modes. At first, we introduce the details of this ultrathin SiO_2 film technique.

2.1 Ultrathin SiO_2 film technique for Ge and Si QD formation

Si(111) or Si(001) substrates were introduced into an ultrahigh vacuum chamber at the base pressure of ~1×10^{-8} Pa. Clean Si surfaces were obtained by flashing at 1250°C or by forming Si buffer layers. The clean Si surfaces were oxidized at ~ 600°C for 10 min at the oxygen pressure of 2×10^{-4} Pa to form ultrathin SiO_2 films (<1 nm) (Matsudo et al., 2002). Then, Si or Ge was deposited on the ultrathin SiO_2 films at 500°C using an electron beam evaporator or a Knudsen cell to form spherical Si or Ge QDs with an ultrahigh density of ~ 2×10^{12} cm^{-2}. Scanning tunneling microscope (STM) experiments were conducted at the sample bias voltage of 3-6 V and tunnelling current of 0.2 nA at room temperature (RT) using sharp chemically treated W tips (Nakamura et al., 1999).

Figures 1(a) and (b) show STM image and reflection high energy diffraction (RHEED) pattern of the ultrathin SiO_2 film, respectively. STM image describes amorphous surface.

STM images of Si and Ge QDs were shown in Figs. 2(a) and 2(b), which were formed by deposition of 8-monolayer (ML) Si and 5-ML Ge, respectively, at 500°C. Spherical QDs of ~5 nm in diameter were grown on Si substrates with an ultrahigh density of ~ 2×10^{12} cm^{-2}. RHEED pattern revealed that at deposition temperature higher than ~450°C, QDs were epitaxially grown with the same crystal orientation of the Si substrates as shown in Figs. 2(c). On the other hand, at deposition temperature lower than ~400°C, RHEED of QDs exhibited Debye ring pattern indicating that QDs were non-epitaxially grown on Si substrates.

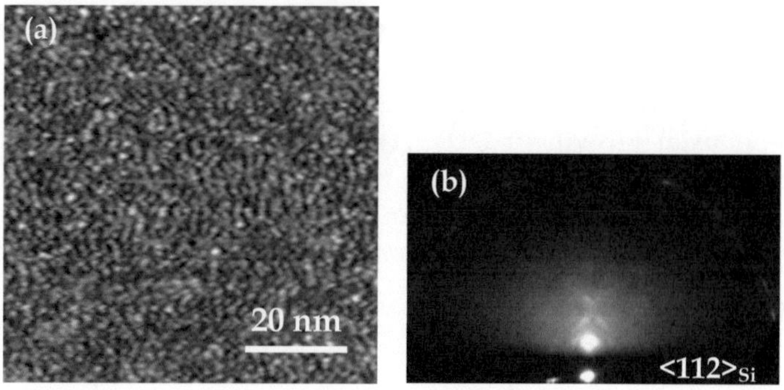

Fig. 1. (a) STM image and (b) RHEED pattern of ultrathin SiO_2 film on Si (111) substrate.

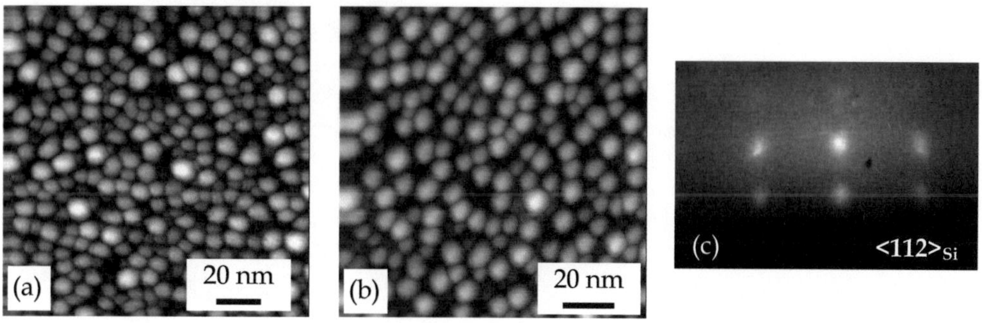

Fig. 2. STM images of Si (a) and Ge QDs (b) on Si (111) substrates formed by deposition of 8-ML Si and 5-ML Ge. (c) RHEED pattern of Ge QDs shown in (b).

2.2 Mechanism of ultrahigh density QD formation by ultrathin SiO_2 film technique

Mechanism of ultrahigh density QD formation (Shklyaev et al., 2000; Shklyaev & Ichikawa, 2001; Nakamura et al., 2004) is illustrated in Fig. 3. At the first stage of Si or Ge deposition on the ultrathin SiO_2 films, deposited atoms diffuse on the surfaces and react with the ultrathin SiO_2 films after certain lifetime, τ through the following reactions,

$$Si+SiO_2 \rightarrow 2SiO\uparrow, \text{ or } Ge+SiO_2 \rightarrow SiO\uparrow+GeO\uparrow. \tag{1}$$

As a result, at reaction sites, the nanometer-sized windows (nanowindows) are formed in the ultrathin SiO_2 films. The distance between nanowindows is determined by diffusion length of deposited atoms until they react with the ultrathin SiO_2 films, $\sim \sqrt{D\tau}$, where D is the diffusion coefficient. In the case of Si and Ge, this value is approximately 10 nm resulting in the ultrahigh density (~10^{12} cm^{-2}). Nanowindows are functioned as trap sites for deposited atoms due to the dangling bonds. In other words, the chemical potential of deposited atoms at the nanowindow sites is smaller than that on the ultrathin SiO_2 film. Then, nanowindows work as nucleation sites for QD growth resulting in the formation of ultrahigh density spherical QDs. Despite the existence of the ultrathin SiO_2 films, epitaxial growth of our QDs is understood in this scenario as shown in Fig. 4. When the deposition temperature is high (>~450°C), nanowindows form sufficiently through reaction (1). As a result, the formed QDs contact with Si substrates through nanowindows leading to the epitaxial growth. On the contrary, when the deposition temperature is low (<~400°C), nanowindows do not penetrated into Si substrates due to insufficiency of reaction (1), resulting in non-epitaxial growth of QDs. As for epitaxial QDs, the limited nanocontact between QDs and substrates reduces the strain energy in QDs induced by lattice mismatch. The QDs are elastically strain-relaxed without misfit dislocation due to the small strain energy and spherical shape. This unique formation mechanism brings high quality in QDs.

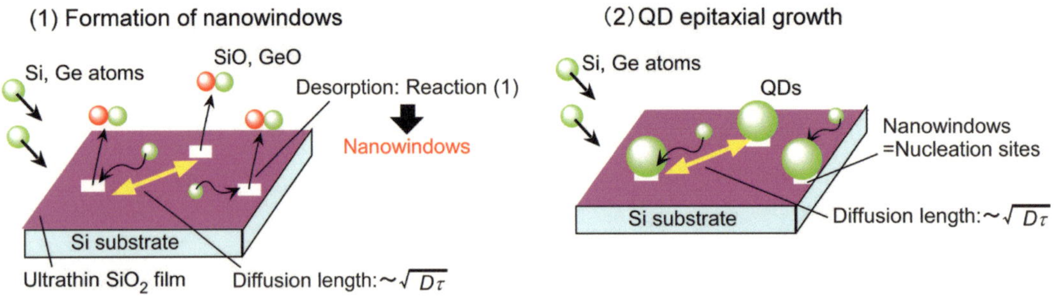

Fig. 3. Illustration of mechanism of ultrahigh density QD formation by ultrathin SiO_2 film technique.

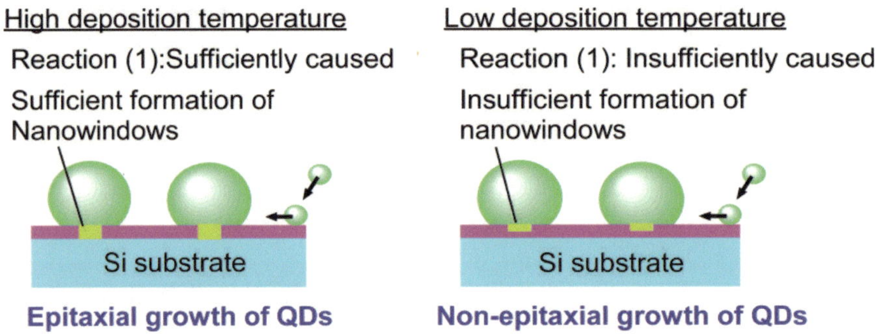

Fig. 4. Schematic of mechanism of epitaxial or non-epitaxial growth of QDs

3. Modified formation technique for QDs of other materials

To apply the aforementioned ultrathin SiO_2 film technique to materials other than Si and Ge, we modified this technique in the following way (Fig. 5) (Nakamura et al., 2006; Nakamura et al., 2009). First, ultrahigh density nanowindows were formed in the ultrathin SiO_2 films by predeposition of Si or Ge. Second, materials A and B were codeposited with flux ratio of 1-x to x to form epitaxial $A_{1-x}B_x$ QDs. In this proposed technique, there are several advantages: (1) Ultrahigh density QDs can be formed because ultrahigh density nanowinows work as nucleation sites. (2) QD size is controllable down to ~1 nm by controlling deposition amount. (3) Tuning flux ratio enables the composition control. (4) Existence of nanowindows allows the epitaxial growth on SiO_2 films. (5) The SiO_2 films prevent diffusion of deposited atoms into Si substrates. (6) The technique can be applied to various materials. (7) Limited contact between QDs and Si substrates enables elastic strain-relaxation in spherical QDs without introduction of misfit dislocation in their heterointerfaces (Nakamura et al., 2010; Nakamura et al, 2011a, Nakamura et al, 2011b).

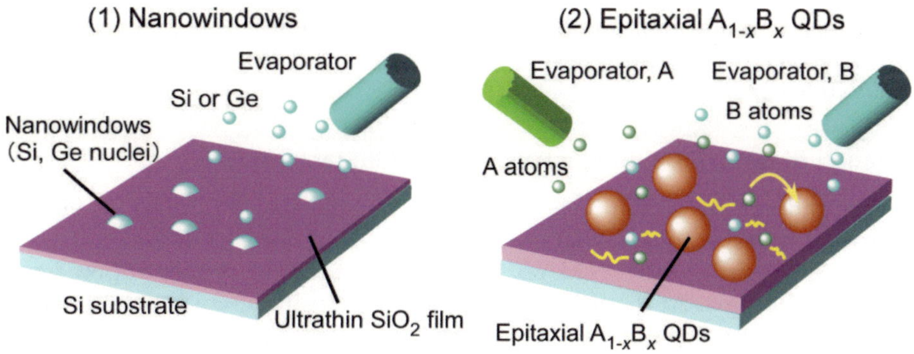

Fig. 5. Idea of the modified ultrathin SiO_2 film technique

Here, we focus on Ge_{1-x} Sn_x which were reported to be direct-transition semiconductor at larger x>~0.12 (Jenkins & Dow, 1987; He & Atwater, 1997). Based on the above idea, we have developed epitaxial growth technique of ultrahigh density Ge_{1-x} Sn_x QDs on Si substrates (Nakamura et al., 2007c) which is a promising material for light emitter. This ultrathin SiO_2 film technique and the physical properties of these QDs are presented.

3.1 Experimental procedure

Samples cut from Si(111) or Si(001) wafers were introduced into an ultrahigh-vacuum chamber with a base pressure of about 1×10^{-8} Pa. The chamber was equipped with a STM, a RHEED apparatus, and two separate Knudsen cells for Ge and Sn deposition. Ultrathin SiO_2 films with thicknesses of ~0.3 nm (Matsudo et al., 2002) were formed by oxidizing the Si surfaces in the chamber at 600°C for 10 min at an oxygen pressure of 2×10^{-4} Pa, after cleaning the Si surfaces by flashing at 1250°C. To form nanowindows in the ultrathin SiO_2 films, we deposited a small amount of Ge on the ultrathin SiO_2 films under various conditions, i.e. a Ge deposition amount of 1-2 ML at temperatures of 500-650°C, a process that is referred to as Ge predeposition. We considered that nanowindow formation by Ge predeposition was needed to initiate contact between the QDs and Si substrates through SiO_2 films to grow epitaxial QDs. The nanowindow formation mechanism has been reported in our papers (Shklyaev et al., 2000; Shklyaev & Ichikawa, 2001; Nakamura et al., 2004). Ge and Sn were codeposited on these ultrathin SiO_2 films with nanowindows at ~100-200°C at the flux ratio of 1-x to x to form $Ge_{1-x}Sn_x$ QDs. This low temperature growth of ~100-200°C was required for the growth of the supersaturated GeSn alloy films (He & Atwater, 1997.; Gurdal et al., 1998, Ragan & Atwater, 2000., de Guevara et al., 2003). Sn atoms might segregate in the samples with large Sn content x, but in the present chapter, the Sn content x in the QDs was described as flux ratio. RHEED patterns were observed using a 15-keV electron beam with an incident direction of <112>$_{Si}$ for Si(111) substrates and <110>$_{Si}$ for Si(001) substrates. Typical STM experiments were conducted at a sample bias voltage, V_S of +2-5 V and a tunneling current, I_T of 0.1 nA at room temperature using chemically-sharpened W tips (Nakamura et al., 1999). STS experiments were performed at V_S of +0.8 V and at I_T of 0.1 nA at RT using chemically-polished PtIr tips cleaned by electron beam heating in UHV chamber. Before STS measurements, in order to remove the surface state contribution in the energy bandgap, we performed atomic hydrogen termination of the QD surfaces by introducing hydrogen molecules up to 2×10^{-4} Pa for 70 min at room temperature after putting the samples in front of heated W filaments. When we observed cross section of the samples using ex-situ high-resolution transmission electron microscopy (HRTEM) and scanning transmission electron microscopy (STEM), the samples were covered with amorphous 20-ML Si films.

3.2 Formation of GeSn QDs using ultrathin SiO_2 film technique

Figures 6(a) and (b) are STM image and RHEED pattern of the ultrathin SiO_2 film on Si(111) substrates, respectively, where nanowindows were formed by Ge predeposition in 2-ML amounts at 650°C. The STM results revealed that Ge predeposition formed Ge nuclei on the SiO_2 films. At the first stage of Ge predeposition, voids were formed in the ultrathin SiO_2 films through reaction (1), and the extra atoms of deposited Ge were trapped at the void sites resulting in the formation of Ge nuclei. We call the Ge nuclei on voids as nanowindows in this case. Ge nuclei has ultrasmall volume due to the volatilization of GeO. Obscure 1×1 diffraction patterns in Fig. 6(b) indicates that the Ge nuclei were crystals epitaxially grown on the substrates through the voids, with the same crystallographic orientations as those of the substrates.

Fig. 6. (a) STM image and (b) RHEED pattern of the ultrathin SiO_2 films with predeposited Ge in 2-ML amounts at temperatures of 650°C.

After the nanowindows were formed on the ultrathin SiO_2 films by 2-ML Ge deposition at 650°C, we codeposited Ge and Sn at 200°C to form $Ge_{0.85}Sn_{0.15}$ QDs as shown in Fig. 7(a). In order to prevent Sn segragation, codeposition temperature should be less than ~200°C . Spherical $Ge_{0.85}Sn_{0.15}$ QDs of ~5 nm in diameter were formed with an ultrahigh density (~2×10^{12} cm^{-2}). RHEED pattern of this sample shows the diffraction pattern revealing the epitaxial growth of QDs with the same orientation as that of Si substrates.

Morphorlogy of $Ge_{0.85}Sn_{0.15}$ QDs was independent of predeposition condition for the nanowindow formation. However, the crystal orientation relationships between QDs and substrates strongly depended on the predeposition condition. We closely observed RHEED patterns of $Ge_{0.85}Sn_{0.15}$ QDs codeposited at 100°C, where the nanowindows were formed under various predeposition conditions. In the case of low predepositoin temperature of 500°C, Debye rings appeared in RHEED patterns demonstrating non-epitaxial growth of QDs as shown in Fig. 7(c), regardless of the amount of predeposited Ge amout (1-2 ML). On the other hand, in the case of high predeposition temperature of 650°C, epitaxial spots were observed in the RHEED patterns as shown in Fig. 7(b). However, for small amounts of Ge predeposition (1 ML), Debye rings were also observed, albeit only slightly, in addition to the spot patterns. Epitaxial spots without Debye rings appeared only in the case of 2-ML Ge predeposition, indicating that the $Ge_{1-x}Sn_x$ QDs were epitaxially grown only if 2-ML Ge was predeposited on the ultrathin SiO_2 films at 650°C for nanowindow formation. The predeposition condition is essential for epitaxial growth of QDs. For Si(001) substrates, the formation of $Ge_{1-x}Sn_x$ QDs using the same methods was confirmed, indicating that this formation technique does not depend on the face orientations of substrates.

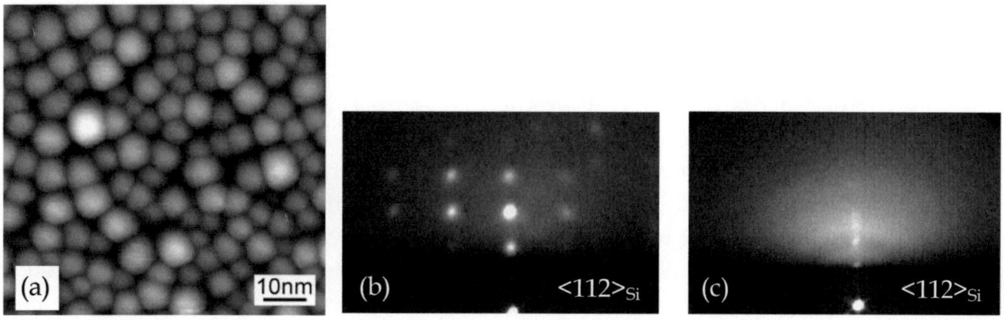

Fig. 7. (a) STM image of 8-ML $Ge_{0.85}Sn_{0.15}$ QDs formed by codeposition of Ge and Sn at 200°C on the ultrathin SiO_2 films with nanowindows. (b, c) RHEED pattern of 8-ML $Ge_{0.85}Sn_{0.15}$ QDs fomred by codeposition at 100°C. 2ML Ge was predeposited on the ultrathin SiO_2 films at 650 (a, b) and 500° C (c).

Reaction (1) occurred only during the Ge predeposition stage and not during the codeposition, because the codeposition temperatures (~100-200°C) were lower than that needed for reaction (1) (>~400°C) (Shklyaev et al., 2000). Excessive Ge atoms, which were not consumed by reaction (1) during predeposition, were used for epitaxial growth of Ge nuclei on the ultrahigh density voids resulting in the formation of ultrahigh density nanowindows, the mechanism of which is the same as that for ultrahigh density Ge QD formation (Shklyaev et al., 2000). The nanowindows worked as nucleation sites for $Ge_{1-x}Sn_x$ QD formation during codeposition, resulting in the epitaxial growth of ultrahigh density QDs. Under other predeposition conditions, (namely Ge amounts less than 2 ML or temperatures less than 650°C), the non-epitaxial growth of $Ge_{1-x}Sn_x$ QDs revealed the insufficiency of reaction (1). Increases in the amount of predeposited Ge or predeposition temperature can cause sufficient nanowindow formation. However, an excessive increase in the amount of predeposited Ge causes the Sn content, x in $Ge_{1-x}Sn_x$, to deviate from the objective value of ~0.15. Furthermore, a substrate temperature larger than 700°C causes the decomposition of SiO_2 films. Therefore, the Ge nucleus formation condition, a key factor for the epitaxial growth of $Ge_{1-x}Sn_x$ QDs, is limited.

The $Ge_{1-x}Sn_x$ QDs were annealed at 500°C for 3 min to improve low crystallinity of low-temperature-grown QDs (200°C). Althogh this annealing process caused only a slight increase in the QD diameter and a slight decrease in the QD density by QD coalescence, QD size and diameter are still similar to those of non-annealed QDs, namely, nanometer-size and ultrahigh density (~10^{12} cm^{-2}). Then, we used this condition for formation of epitaxial $Ge_{1-x}Sn_x$ QDs: predeposition of 2-ML Ge at 650°C, codepositon of Ge and Sn at 200°C, and annealing at 500°C for 3 min. We refer to this condtion as the epitaxial QD growth condition. Figure 8(a) shows a cross-sectional HRTEM image of the 8-ML epitaxial $Ge_{0.85}Sn_{0.15}$ QDs on Si(111). The QDs were fabricated under the above epitaxial QD growth condition. Spherical $Ge_{0.85}Sn_{0.15}$ QDs were formed with clear interfaces on the Si surfaces. The diameter and height of the QDs are ~7 and ~3 nm, respectively. These values were more precise than values measured in the STM images owing to STM tip apex effect. An FFT pattern of the $Ge_{0.85}Sn_{0.15}$ QD area shows a diffraction pattern corresponding to the epitaxial growth on Si(111). An inverse fast Fourier transform (IFFT) was performed through the $1\bar{1}0$ spots in the FFT pattern corresponding with the dashed square area in Fig. 8(a), where the area included the interface between the $Ge_{085}Sn_{0.15}$ QDs and Si substrate. The lattice planes were clearly shown in IFFT image in Fig. 8(b). No differences were found between the numbers of atomic planes on the sides of Si substrates and of QDs near their interfaces in the IFFT images in Fig. 8(b), except for the line deformation. This indicated there were no misfit dislocations at the interfaces between QDs and Si substrates. Furthermore, the line spacing became larger in the upper part of the QD in Fig. 8(b) although the line spacing in the QD was almost the same as that of Si substrates at the interface. Therefore, we considered that some strain in the upper part of QDs relaxed due to the spherical shape of QDs, instead of the formation of misfit dislocation. Figure 8(c) is an STEM image of the same sample with the same electron incident direction used for the HRTEM observation, where it is possible to detect the atomic number difference as the contrast difference. No contrast difference was found within the QDs, indicating no Sn segregation in the QDs within the spatial resolution of this measurement. To estimate the lattice constant of the $Ge_{0.85}Sn_{0.15}$ QDs, the FFT patterns of the $Ge_{0.85}Sn_{0.15}$ QD areas were compared with those of perfect Si substrate areas far from the interface. The lattice constants of the QDs in the in-plane direction (a_i) and growth direction (a_g) were measured from $1\bar{1}0$ and 222 diffraction spots in FFT patterns. For the main $Ge_{0.85}Sn_{0.15}$ QDs, there is almost no difference between a_i and a_g within the experimental errors. This indicates that main QDs exhibit little or no strain. The lattice constants individually measured for these QDs were 0.575±0.03 nm. One might consider that less-strained QDs were grown by the formation of misfit dislocations. However, the HRTEM images exhibited no misfit dislocations in the QDs. This indicates that heteroepitaxial strain is relaxed elastically in spherical QDs due to their sperical shape. Originally, there are two reasons that the present QDs would have small strain energy even if QDs were not be elastically strain-relaxed. One is the ultrasmall QD volumes that can have only small strain emergy. The other is due to the small contact betwen QDs and substrates through nanowindows (~1nm) in the ultrathin SiO_2 films (Nakamura et al., 2007a. Nakamura et al., 2010) unlike the heteropetaixal films or Stranski-Krastanov islands. These nanowindows were difficult to observe in Fig. 8(a). The measured lattice constants of QDs with almost no strain (0.575±0.003 nm) are larger than those of Ge (0.565 nm), which is derived from the alloying of Ge and Sn, rather than from the strain effect in the heterostructures. According to reports on the strain-relieved $Ge_{1-x}Sn_x$ alloy films (He & Atwater, 1997; de Guevara et al., 2003), the lattice constant of the $Ge_{1-x}Sn_x$ QDs gave a Sn concent x of 0.1-0.13. The Sn content value was smaller than the Sn flux ratio of 0.15 used during codeposition. This can be explained by considering the existence of Ge nuclei formed by Ge predeposition. In the case of 8-ML $Ge_{1-x}Sn_x$ QDs, 6.8-ML Ge and 1.2-ML Sn were deposited during the codeposition process. Therefore, considering an amount of Ge nuclei less than 2 ML, the content x ranges from 0.12 to 0.15, which agreed approximately with the experimental results. The HRTEM observation revealed that QDs were elastically strain-relaxed without misfit dislocations and that the Sn content in $Ge_{1-x}Sn_x$ alloy is close to 0.15. This high x value and high crystal quality can be achieved only in the present QDs. The high x value in our $Ge_{1-x}Sn_x$ QDs is found to be astounding by comparing with that of reported $Ge_{1-x}Sn_x$ alloy films. This may be due to the aforementioned elastic strain-relaxation in nanometer-sized spherical QDs.

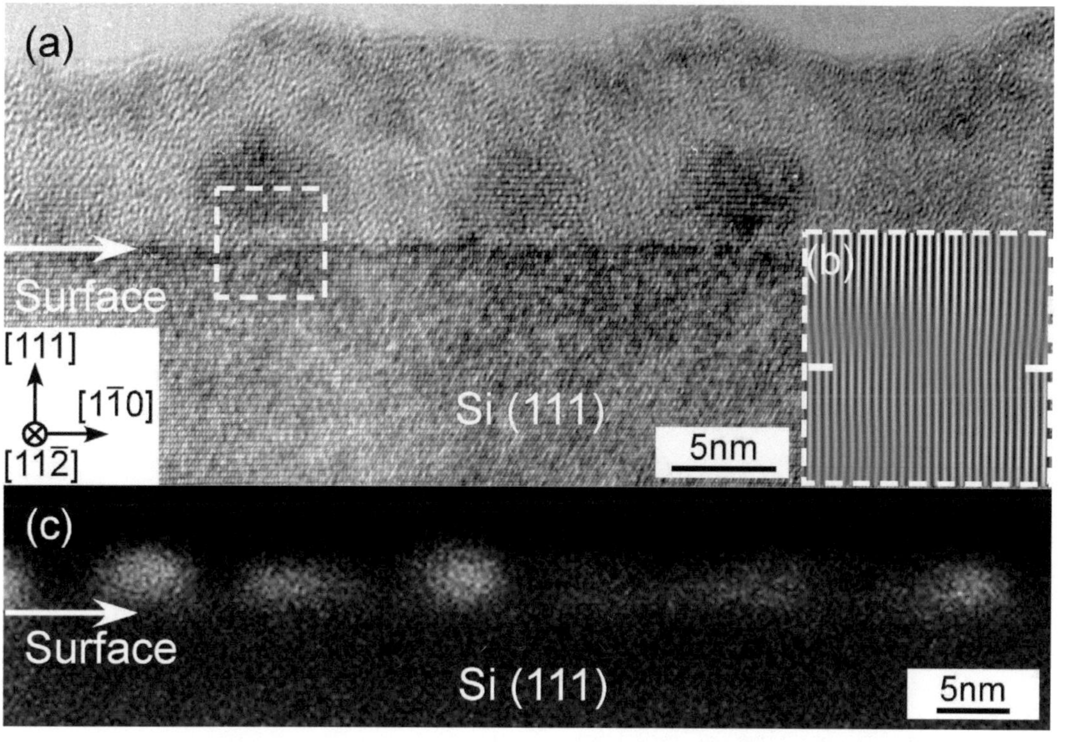

Fig. 8. (a) Cross sectional HRTEM image and (c) STEM image of epitaxial 8-ML $Ge_{0.85}Sn_{0.15}$ QDs on Si(111) formed under the epitaxial QD growth condition. (b) IFFT image of dotted squre area in (a). Two white lines in (b) indicat interface between QDs and substrate.

3.3 Electronic properties

We investigated electronic sates of $Ge_{1-x}Sn_x$ QDs individually using STS measurement (Nakamura et al., 2007b.). The QDs we measured are prepared by the hydrogen-termination of epitaxial $Ge_{0.85}Sn_{15}$ QDs formed under the epitaxial QD growth condition. The differential conductances, dI/dV, in Figs. 9(d)-(f) were obtained by STS measurements on the top of the hydrogen-terminated epitaxial $Ge_{0.85}Sn_{15}$ QDs indicated by the arrows in STM images in Figs. 9(a)-(c), respectively. Peaks in the conduction (p+) and valence (p-) band regions were observed in these spectra, though the p- peaks were sometimes not discernible. Difficulties of observing peaks in the valance band have been reported for other semiconductors (Grandidier, 2004). Figure 9 demonstrates that the p+ peak up-shifted with the decrease in the QD size. We also found various peak shapes, sharp and step-like peaks, as shown in Figs. 9(d) and (f). The peaks were broader for flattened QDs than those for spherical QDs.

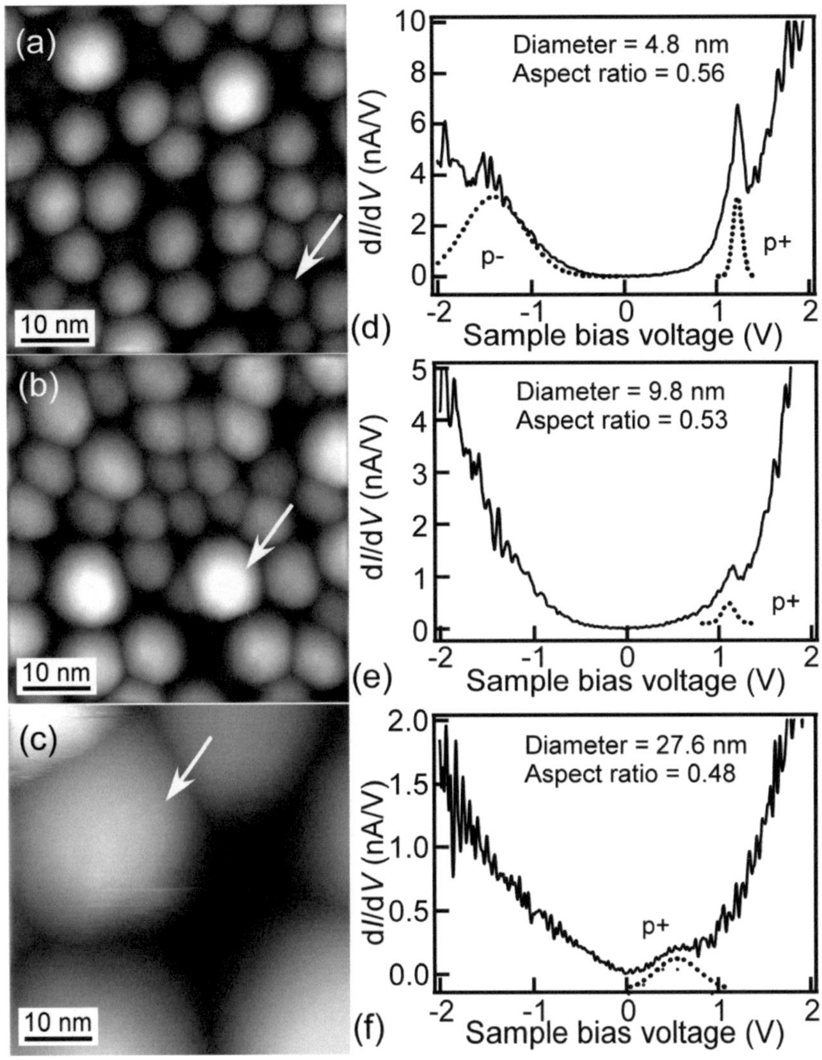

Fig. 9. (a), (b), (c) STM images of annealed $Ge_{0.85}Sn_{0.15}$ QDs with various deposition amount of 4-24 ML. (d), (e), (f) STS results corresponding to QDs indicated by arrows in (a), (b) and (f), respectively.

We measured the peak position and its width by deconvolution of the spectra using Gaussian functions with a standard deviation, σ as shown by dashed lines in Figs. 9 (d)-(f). We defined the CBM (VBM) by subtracting (adding) 2σ from (to) the peak position of p+ (p-). Figure 10(a) show the dependences of CBM and VBM on the diameter, d of the QDs. It is clear that the absolute values of both VBM and CBM up-shifted by about 0.5 eV with the decrease in the QD diameter from 35 to 4 nm. The QD-diameter dependence of the energy bandgap, defined as CBM-VBM, is shown in Fig. 10(b). The energy bandgap increased with a decrease in QD size. An interesting thing is the origin of the peaks in the STS spectra for the QDs. The peaks in the region of the energy bandgap may originate from the surface states. In the present case, however, we measured the hydrogen-terminated QDs, where the surface states were removed in the region of the energy bandgap (Boland, 1991). Moreover, the peak positions shifted depending on the QD size. These results ruled out the possibility that the peaks originated from the surface states. Therefore, we considered that the peaks corresponded with the quantum levels derived from the quantum-confinement effect in the $Ge_{1-x}Sn_x$ QDs.

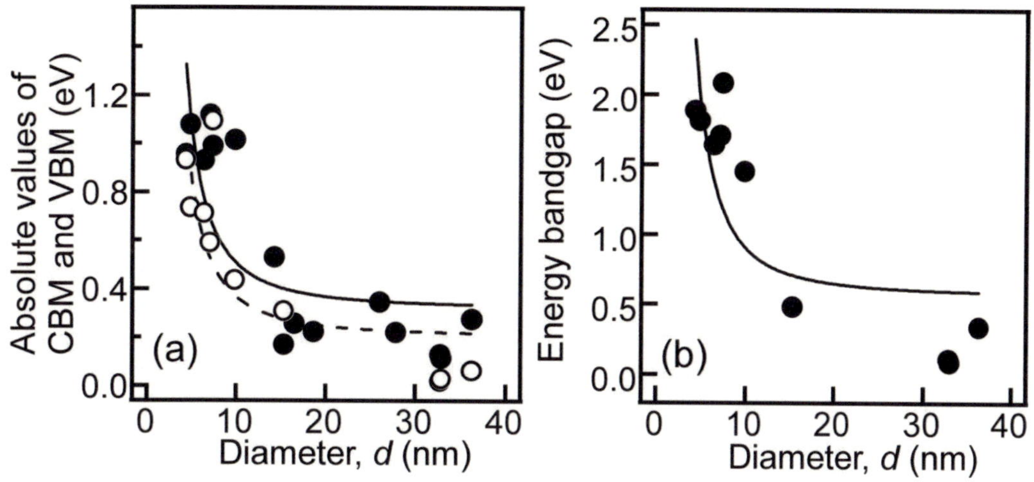

Fig. 10. (a) CBM (circles) and VBM (cricles) estimated from STS results of $Ge_{0.85}Sn_{0.15}$ QDs. (b) Energy bandgap calculated from CBM -VBM .

We investigated the dependence of the peak width of p+, defined as 2σ on the QD aspect ratio (=height to diameter). Figure 11(a) shows the strong dependence of the peak width on the QD aspect ratio. The peaks became broader as the QDs changed from spherical to flat ones (quantum well for an extremely small aspect ratio). In general, the density of states (DOS) in QDs changes gradually as a function of the dimensions of the QD (Grandidier, 2004). As QDs become flattening, the DOS of QDs gradually changes from sharp peak

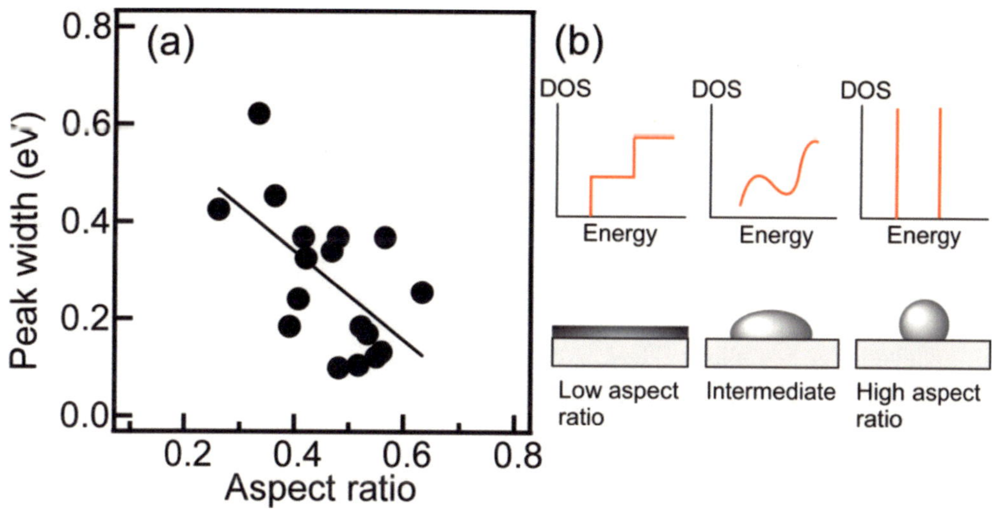

Fig. 11. (a) Dependence of peak width in STS on aspect ratio of QDs. (b) Illustration of density of states depending the QD shape.

structures for 0-dimensional (0-D) QDs to step-wise structures with electron and hole subbands for 2-D structures, through broad peak structures for states that are quasi-0-D QDs. As illustrated in Fig. 11(b), the dependence of the peak width on the QD aspect ratio can be explained by the gradual emergence of 2 dimensionality in DOS of quasi-0-D QDs as the QDs became flatter.

As mensioned in the 3.2 section, there were no differences between the lattice constants in in-plane and growth directions estimated by analysis of cross sectional HRTEM images, which indicated strain-free QDs (Nakamura et al., 2007c). Furthermore, STEM contrast in QDs was uniform. From these results, we considered that within this measurement accuracy, the present QDs were uniform in alloying and strain-free $Ge_{1-x}Sn_x$ QDs with x of 0.1-0.13. In the present situation, the strain-free QDs are sandwiched between vacuum and SiO_2 films. There are nanowindows between the SiO_2 films and the QDs, but they are ultrasmall (<~1nm), so that the carriers can hardly penetrate into the substrates (Nakamura et al., 2007a). The SiO_2 films work as the high barrier for the carriers in the QDs. Therefore, the hard wall square potential model is considered to be reasonable. For the quantum-confinement effect for QD diameter, d, the absolute value of CBM (VBM) for a confined structure, E_{confined} is described as

$$E_{\text{confined}}(d) = E_{\text{bulk}} + \frac{\hbar^2 \pi^2}{2\mu(d/2)^2} \tag{2}$$

where E_{bulk}, is the absolute value of CBM (VBM) for the bulk structure, and μ is the effective mass of the electron (hole). The solid (dashed) lines in Figs. 10(a) are best-fitted curves, where E_{bulk} for the absolute values of CBM (VBM) with respect to the Fermi level, and μ for the electron (hole) were adjusted to 0.32±0.09 eV (0.21±0.11 eV) and $0.080 \pm 0.018 m_0$ ($0.092 \pm 0.026\ m_0$), respectively. The theoretical effective mass of an electron, m_e is reported to be $\sim 0.1 m_0$ (Jenkins & Dow, 1987) and that of the hole is expected to be similar to the hole effective mass for Ge, m_h ($\sim 0.076\ m_0$) because the valence band structure near the Γ point is insensitive to the alloying of Ge and Sn (Jenkins & Dow, 1987). From the rough consistency between these values and the fitted μ values in Figs. 10(a), we found that our experimental result of the size dependence agreed with the quantum-confinement effect model for the 0-D QDs demonstrating that the present QDs were quasi-0-D structures.

The size dependence of energy bandgap due to the quantum-confinement effect can also be written by Eq. (2) when E_{bulk} and μ are the bulk energy bandgap and the reduced electron-hole mass, respectively, for $Ge_{1-x}Sn_x$ alloys. The solid line in Fig. 10(b) is a best-fitted curve with Eq. (2) where the E_{bulk} and μ were adjusted to 0.56±0.2 eV and $0.044 \pm 0.012 m_0$, respectively. This μ value agreed with the calculated one ($\sim 0.043\ m_0$) from m_e ($\sim 0.1\ m_0$) and m_h ($\sim 0.076\ m_0$). According to the relationship between energy bandgap and Sn content, x, for strain-relieved $Ge_{1-x}Sn_x$ alloys (He & Atwater, 1997; de Guevara et al., 2004), the energy bandgap was inferred as ~0.4-0.5 eV for x of 0.1-0.13. This value was roughly consistent with the fitted value E_{bulk} of 0.56±0.2 eV.

3.4 Optical properties

We formed Si capping layers on GeSn QDs by successive Si deposition at 400°C after 3 nm Si deposition at 200°C to fabricate the structures of Si/$Ge_{1-x}Sn_x$ QDs/Si(001). For PL measurement of the Si/$Ge_{1-x}Sn_x$ QDs/Si(001) structures, we used a HeCd laser with a 325 nm wavelength and InGaAs photomultiplier detector. Figures 12(a) and (b) are STM image and RHEED pattern of the 40 nm Si capping layer formed on the $Ge_{0.85}Sn_{0.15}$ QDs epitaxially grown on Si (001) substrates that is Si/$Ge_{0.85}Sn_{0.15}$ QDs/Si . As mentioned in 3.2 section, spherical QDs were epitaxially grown on Si(001) with an ultrahigh density of ~10^{12} cm^{-2} under the epitaxial QD growth condition as is the case with Si (111) substrates. They show the epitaxial Si capping layer formed with the same crystallographic orientation as those of Si (001) substrates and of the underlying QDs and substrates.

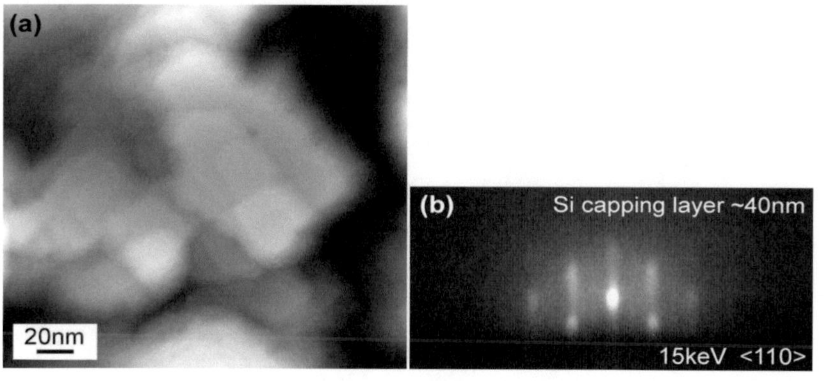

Fig. 12. (a) STM image and (b) RHEED pattern of Si/$Ge_{0.85}Sn_{0.15}$ QDs/Si(001) structure

To investigate the quantum size effect, we formed epitaxial $Ge_{0.85}Sn_{0.15}$ QDs with various QD diameters *d* (7, 9, and 13 nm) under the epitaxail QD growth condition by changing the deposition amounts (5, 8, and 14 ML) and covered these QDs with 60 nm Si capping layers. RHEED patterns of these Si capping layers indicated all Si capping layers were epitaxially grown. With a decrease in the size of the underlying QDs, the RHEED patterns become streakier, indicating the Si capping layer is becoming flatter. We measured PL spectra of these Si-capped $Ge_{0.85}Sn_{0.15}$ QDs with various QD sizes at 5 K as shown in Fig. 13. The PL peak appeared near 0.8 eV in all samples and the peak position was almost independent of the QD size indicating there was no quantum size effect. On the other hand, the PL intensity at the peak increased with a decrease in the QD size. The inset in Fig. 13 shows PL results of as-grown and post-annealed samples of the 60-nm-Si-capped Ge QD structures (dotted lines) that is Si/Ge QDs/Si sturcutres, for reference (Shklyaev et al., 2006.; Ichikawa et al., 2008). The inset shows these as-grown $Ge_{0.85}Sn_{0.15}$ QD samples have PL intensity about 10 times higher than that of as-grown Ge QD samples. The high intensity of these as-grown samples is comparable to that of the high-temperature-post-annealed Ge QD samples (900°C for 30 min).

The experimental result of the increase in the PL intensity with a decrease in QD size as shown in Fig. 13 is explained by two possible mechanisms. One is that these PL results come from the optical transition between the quantum levels in the QDs, where the oscillator strength is enhanced owing to the quantum confinement effect caused by the QD size decrease. The second is the reduction of the non-radiative recombination centers in the Si capping layer due to flatter Si capping layers with a decrease in the size of the underlying QDs. The experimental result with no quantum size effect in PL rules out the light emission mechanism from the QDs. In comparison with the similar PL of Si-capped Ge QDs

(Shklyaev et al., 2006; Shklyaev et al., 2007; Ichikawa et al., 2008), we consider the PL near ~1.5 μm of the present samples presumably originated from the radiative electronic states in the Si capping layers. This also reveals that the PL intensity enhancement due to the QD size decrease is caused by the latter mechanism; that is, the reduction of non-radiative recombination centers in the Si capping layer.

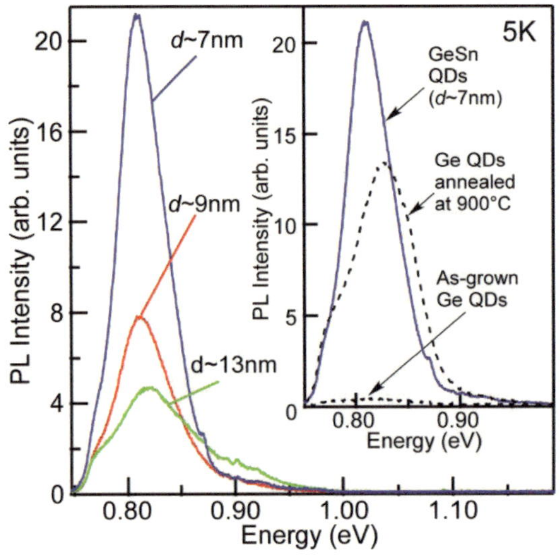

Fig. 13. PL spectra of Si/$Ge_{0.85}Sn_{0.15}$ QDs/Si(001) structures with QD diameter d of 7, 9 , and 13 nm. In inset, PL spectra of Si/Ge QDs/Si(001) structure before and after annealing at 900 are shown for reference.

To investigate the influence of Ge-Sn alloying in QDs on the high PL intensity, we measured the dependence of the PL spectrum on the Sn content x in the $Ge_{1-x}Sn_x$ QDs. 40 nm thick Si capping layers formed on 9 nm diameter (8 ML) $Ge_{1-x}Sn_x$ QDs with values for x of 0, 0.07, 0.15, 0.25, and 0.5. To distinguish between the alloying effect and Sn segregation effect, we also formed 6.8 ML Ge QDs and then deposited 1.2 ML Sn at 200°C. These separately-deposition (SD) dots were covered with 40 nm Si capping layers. For these SD samples, the deposition amount was the same (8 ML) and the ratio of Ge and Sn deposition amounts was 0.85 to 0.15. At small x of 0 to 0.15, RHEED patterns were streaky as shown in Fig. 12(b), indicating that Si capping layers were flat. At x=0.25, the RHEED pattern in Fig. 14(a) shows obscure 1/4 spots indicated by lines in addition to streaky Si diffraction patterns. The 1/4 spots are considered to come from the c(4×4) Sn-absorbed Si(001) surfaces (Baski et al., 1991; Lyman & Bedzyk, 1997), revealing that a small number of Sn atoms segregated to the surface during the formation of the Si capping layer. Both the QD samples with x = 0.5 and the SD samples show spotty RHEED patterns for the Si capping layers as shown in Fig. 14(b) indicating rough surface of Si capping layer. In the samples including the QDs with large x = 0.5 and the SDs, Sn segregation was presumably caused because these samples had large Sn concentrations at the interfaces between QDs and Si capping layers. PL spectra of these Si-capped QDs were measured as shown in Fig. 14(c). The integrated PL peak intensity in the range of 0.75-1.0 eV is also shown in Fig. 14(d). These results reveal that the integrated PL peak intensity has a maximum near x=0.25. The SD samples show the lower PL intensity (open circles) compared with that of the $Ge_{0.85}Sn_{0.15}$ QD sample with the same Sn content x.

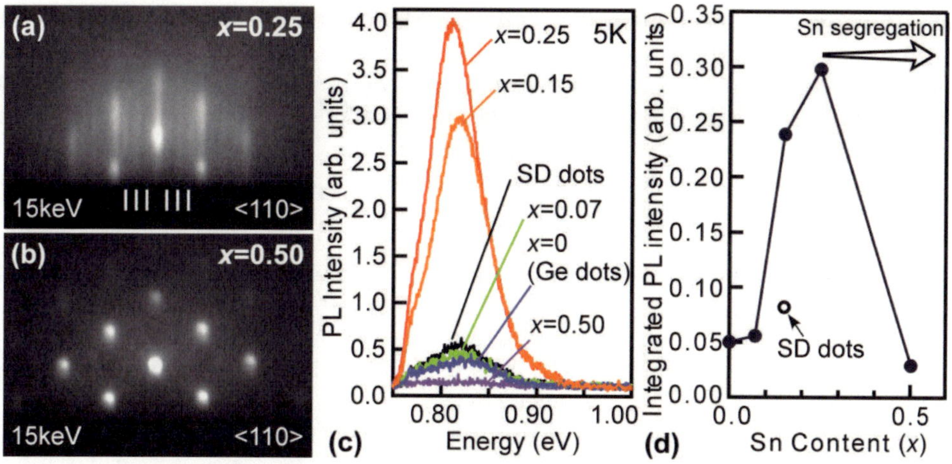

Fig. 14. (a), (b) RHEED pattern of $Si/Ge_{1-x}Sn_x$ QDs/Si(001) structures with x of 0.25 (a) and 0.5 (b). (c) PL spectra of $Si/Ge_{1-x}Sn_x$ QDs/Si(001) structures with various x. (d) Dependence of integrated PL intensity on Sn content x.

We investigated the temperature dependence of the PL peak intensity and PL spectrum as show in Fig. 15. The PL intensity in Fig. 15 bgan to decrease at a temperature higher than 40 K. The integrated PL peak intensity was fitted with $a/(1+b\exp(-E_a/kT))$, where E_a is an activation energy for thermal quenching of the PL, a and b are coefficients, k is the Boltzmann constant, and T is the measurement temperature. As a result, E_a was adjusted to ~22 meV, which is similar to that for the Si-capped Ge QD structures (Ichikawa et al., 2008).

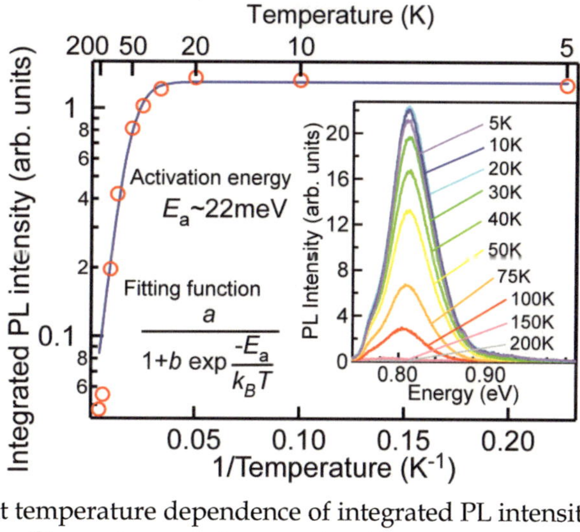

Fig. 15. Measurement temperature dependence of integrated PL intensity of $Si/Ge_{0.85}Sn_{0.15}$ QDs/Si(001) structure. PL spectra was shown in inset.

The PL intensity in the present structures is about ten times higher than that of as-grown Si-capped Ge QD structures. The only difference is the existence of Sn atoms. There are two possible effects: Sn segregation and the Ge-Sn alloying effect. Sn segregation could enhance the formation of the radiative electronic states or remove non-radiative recombination defects. However, in both the $Ge_{0.5}Sn_{0.5}$ QD sample and the SD sample where Sn atoms segregated into the Si capping layer, the PL intensity reduced as shown in Fig. 14(c). This indicates Sn segregation introduces non-radiative recombination defect centers rather than enhancing the PL intensity. On the other hand, Ge-Sn alloying results in a lattice mismatch between the QDs and Si capping layer larger than that for Ge QDs. In the case of Si-capped Ge QDs, high-temperature-post-annealing can cause the relaxation of strain between Si capping layers and Ge, which plays a role in the formation of radiative electronic states in Si capping layers (Shklyaev et al., 2006). When we adopt this strain-driven formation mechanism of the radiative electronic states, the enhancement of PL intensity can be explained by larger strain in Si-capped $Ge_{1-x}Sn_x$ QDs. At smaller $x<0.25$, where Sn segregation was negligible (Nakamura et al., 2007c), the PL intensity increased monotonically with an increase in x; that is, an increase in the lattice mismatch. This experimental result supports our proposed mechanism: GeSn QDs work as nanometer-size-controlled stressors to introduce radiative electronic states in Si capping layers. Also, the consistencey of the activation energy for thermal quenching between Ge and GeSn QD samples can be reasonable in this model of radiative electronic states in Si capping layer. In the case of Si bulk with radiative defects formed by other techniques such as long-duration or high-temperature-annealing, the PL intensity was reported to be enhanced by introducing strain (Kveder et al.,1995; Leoni et al., 2004). Although the origin of radiative electronic states is still open to question, the strain-driven formation of the radiative electronic states in the Si layer is a plausible mechanism for the enhancement of PL intensity in the present samples.

3.5 Optical properties of modified structures

We investigated hydrogen-plasma treatment effect on PL spectra of Si/$Ge_{1-x}Sn_x$ QDs/Si(001) structures because non-radiative recombination centers can be terminated by hydrogen atoms in general. First, we measured PL spectra of Si/Ge QDs/Si(001) structure (the case of x=0) as shown in Fig.16. Hydrogen-plasma treatment was performed at 280°C. This result revealed shape and position of PL peaks were changed by hydrogen-plasma treatment. This demonstrated that the hydrogen-plasma treatment affected the radiative electronic states in Si capping layer. Although the origin of this change is not elucidated yet, there is a possilbity of enhancing the light emitting efficiency and improving thermal properties of Si/GeSn QDs/Si structures by optimizing the treatment condition.

Next, to change barrier materials, we fabricated SiO_2/GeSn QDs/Si sturctures. SiO_2 capping layers of ~70 nm in thickness were formed by Si deposition at the O_2 partial pressure of 2×10^{-4} Pa as shown by RHEED pattern in Fig.17. First 7 nm SiO_2 layer was formed at RT followd by formation of 63 nm SiO_2 layers at 400°C.

We measured PL of SiO_2/8-ML $Ge_{0.85}Sn_{0.15}$ QDs/Si(001) at 5K. There is no PL signal in 0.8 eV range. In order to improve quality of SiO_2, rapid thermal annealing at 800°C for 5 s was peformed, PL of which smaple is shown in Fig. 18(a). The Broad peak was observed in the range of 0.8-1.0 eV. However, sharp structure in PL spectra did not appear. We also measured PL spectra of this sample in visible region as shown in Fig. 18 (b). The strong visible PL was observed, which can be related to the defects in SiO_2 layer. This PL peak has strong light emitting efficency even at RT. Therefore, there is a possibility that 0.8 eV PL was weakend due to the existnece of the defects in SiO_2 causing visible PL with high light emitting efficiency. This indicates that there is a room of developing SiO_2/GeSn QDs/Si structures with high light emitting efficiency near 0.8 eV by improving SiO_2 capping layer.

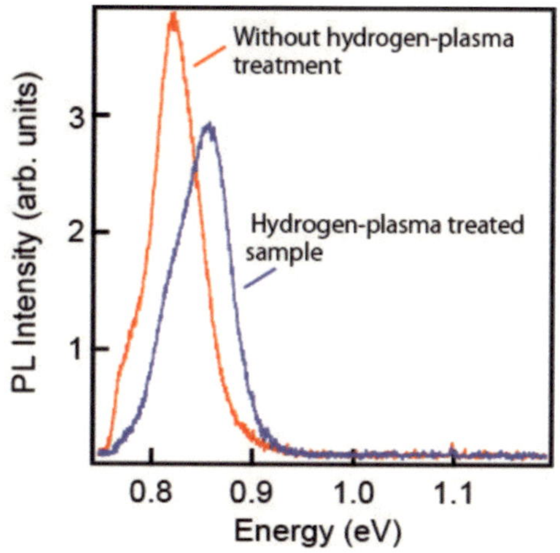

Fig. 16. PL spectra of Si/Ge QDs/Si(001) structures with and without hydrogen-plasma treatment.

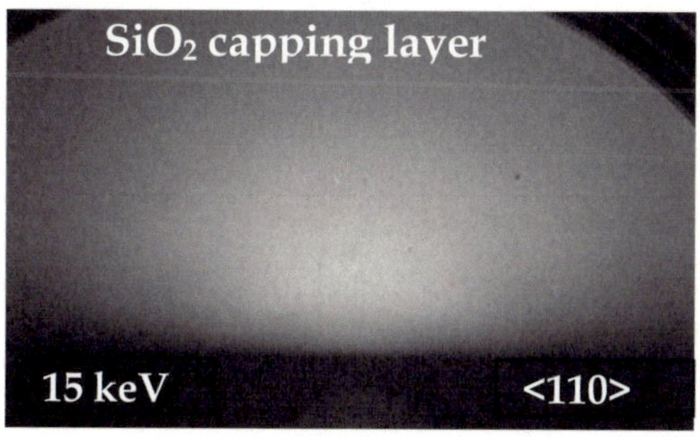

Fig. 17. RHEED pattern of SiO_2 layer formed on 8-ML $Ge_{0.85}Sn_{0.15}$ QDs/Si(001)

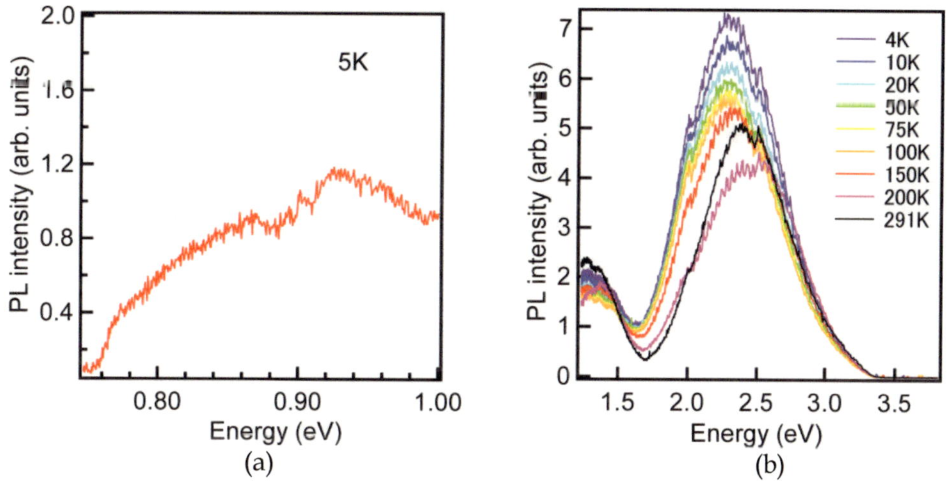

Fig. 18. PL spectra arounnd 0.9 (a) and 2 eV (b) of SiO_2/8-ML $Ge_{0.85}Sn_{0.15}$ QDs/Si(001) structure.

4. Conclusion

We developed the epitaxial growth technique of ultrahigh density GeSn QDs on Si substrates using the ultrathin SiO_2 films. For the epitaxial growth , nanowindows were formed on the ultrathin SiO_2 films by predeposition of 2-ML Ge at 650°C, which condition is key factor. Annealing at 500°C for 3 min improved low crystallinity of low-temperature-grown QDs, which did not cause Sn segregation. The GeSn QDs exhibited the quantum confinement effect which was observed by STS. STS results revealed that the peak width of the quantum levels in QDs depended on the QD shape, which can be explained by electronic state change from the quasi-0-dimensional to quasi-2-dimensional ones. Strong PL was observed in Si/GeSn QDs/Si near 0.8 eV. This origin can be related to radiative electronic states in Si capping layer, which formation strongly depended on the strain-states near the interfaces between QDs and Si capping layers. The radiative electronic states can be changed by hydrogen-plasma treatment. From SiO_2-capped GeSn QDs, featureless PL appeared at around 0.8 eV. Strong visible PL was also observed even at RT which is related to SiO_2 defects. These results indicate material design based on GeSn QDs/Si structures exhibites a possiblity to develop the Si-based light emitting materials near 0.8 eV with high efficiency.

5. Acknowledgment

We thank Prof. N. Tanaka, and Dr. S. P. Cho for TEM and STEM observation.

6. References

Baski, A. A.; Quate, C. F. & Nogami, J. (1991). Tin-induced reconstructions of the Si(100) surface. *Phys. Rev. B*, 44, 11167-11177

Boland, J.J. (1991). Evidence of pairing and its role in the recombinative desorption of hydrogen from the Si(100)-2 x 1 surface. *Phys. Rev. Lett.*, 67, 1539-1542

Cullis, A. G. & Canham, L. T. (1991). Visible light emission due to quantum size effects in highly porous crystalline silicon. *Nature*, 353 335-338

de Guevara, H. P. L., Rodríguez, A. G., Navarro-Contreras, H. & Vidal M. A. (2003). $Ge_{1-x}Sn_x$ alloys pseudomorphically grown on Ge(001). *Appl. Phys. Lett.*, 83, 4942-4944

de Guevara, H. P. L.; Rodríguez, A. G.; Navarro-Contreras, H. &Vidal M. A. (2004). Determination of the optical energy gap of $Ge_{1-x}Sn_x$ alloys with $0<x<0.14$. *Appl. Phys. Lett.*, 84, 4532-4534

Eaglesham, D. J. & Cerullo, M. (1990). Dislocation-free Stranski-Krastanow growth of Ge on Si(100). *Phys. Rev. Lett.* 64, 1943-1946

Grandidier B. (2004). Scanning tunnelling spectroscopy of low-dimensional semiconductor systems. *J. Phys.: Condens. Matter*, 16, S161-S170

Gurdal, O.; Desjardins, P.; Carlsson, J. R. A.; Taylor, N.; Radamson, H. H.; Sundgren, J.E. & Greene, J. E. (1998). Low-temperature growth and critical epitaxial thicknesses of fully strained metastable $Ge_{1-x}Sn_x$ ($x\leq0.26$) alloys on Ge(001) 2×1. *J. Appl. Phys.*, 83, 162-170

He, G. & Atwater H. A. (1997). Interband Transitions in Sn_xGe_{1-x} Alloys. *Phys. Rev. Lett.*, 79, 1937-1940

Ichikawa, M.; Uchida S.; Shklyaev A. A.; Nakamura Y.; Cho S. P. & Tanaka N. (2008). Characterization of semiconductor nanostructures formed by using ultrathin Si oxide technology. Appl. Surf. Sci. 255, 669-671

Jenkins, D. W. & Dow, J. D. (1987). Electronic properties of metastable Ge_xSn_{1-x} alloys. *Phys. Rev. B*, 36, 7994-8000

Kveder, V. V.; Steinman, E. A.; Shevchenko, S. A. & Grimmeiss H. G. (1995). Dislocation-related electroluminescence at room temperature in plastically deformed silicon. *Phys. Rev. B*, 51, 10520-10526

Leoni E.; Binetti, S.; Pichaud, B. & Pizzini S. (2004). Dislocation luminescence in plastically deformed silicon crystals: effect of dislocation intersection and oxygen decoration. *Eur. Phys. J. Appl. Phys.*, 27, 123–127

Lyman, P. F. & Bedzyk, M. J. (1997). Local structure of Sn/Si(001) surface phases. *Surf. Sci.*, 371, 307-315

Maeda, Y. (1995). Visible photoluminescence from nanocrystallite Ge embedded in a glassy $Si0_2$ matrix: Evidence in support of the quantum-confinement mechanism. *Phys. Rev. B*, 51, 1658-1670

Nakamura, Y.; Mera, Y. & Maeda, K. (1999). A reproducible method to fabricate atomically sharp tips for scanning tunneling microscopy. *Rev. Sci. Instrum.*, 70, 3373-3376

Nakamura, Y.; Nagadomi, Y.; Sugie, K.; Miyata, N. & Ichikawa, M. (2004). Formation of ultrahigh density Ge nanodots on oxidized Ge/Si(111) surfaces. *J. Appl. Phys.*, 95, 5014-5018

Nakamura, Y.; Watanabe, K.; Fukuzawa, Y. & Ichikawa, M. (2005). Observation of the quantum-confinement effect in individual Ge nanocrystals on oxidized Si substrates using scanning tunneling spectroscopy. *Appl. Phys. Lett.*, 87, 133119-1-3

Nakamura, Y.; Nagadomi, Y. Cho, S. P.; Tanaka, N. & Ichikawa, M. (2006). Formation of ultrahigh density and ultrasmall coherent β-$FeSi_2$ nanodots on Si(111) substrates using Si and Fe codeposition method. *J. Appl. Phys.*, 100, 044313-1-5

Nakamura, Y.; Ichikawa, M.; Watanabe, K. & Hatsugai, Y. (2007a). Quantum fluctuation of tunneling current in individual Ge quantum dots induced by a single-electron transfer. *Appl. Phys. Lett.*, 90, 153104-1-3

Nakamura, Y.; Masada, A. & Ichikawa, M. (2007b). Quantum-confinement effect in individual $Ge_{1-x}Sn_x$ quantum dots on Si(111) substrates covered with ultrathin SiO_2 films using scanning tunneling spectroscopy. *Appl. Phys. Lett.*, 91, 013109-1-3

Nakamura, Y.; Masada A.; Cho, S. P.; Tanaka, N. & Ichikawa M. (2007c). Epitaxial growth of ultrahigh density $Ge_{1-x}Sn_x$ quantum dots on Si(111) substrates by codeposition of Ge and Sn on ultrathin SiO_2 films. *J. Appl. Phys.*, 102, 124302-1-6

Nakamura, Y.; Sugimoto, T. & Ichikawa, M. (2009). Formation and optical properties of GaSb quantum dots epitaxially grown on Si substrates using an ultrathin SiO_2 film technique. *J. Appl. Phys.*, 105, 014308-1-4

Nakamura, Y.; Murayama A.; Watanabe R.; Iyoda T. & Ichikawa M. (2010). Self-organized formation and self-repair of a two-dimensional nanoarray of Ge quantum dots epitaxially grown on ultrathin SiO_2-covered Si substrates. *Nanotechnology*, 21, 095305-1-5

Nakamura, Y.; Miwa, T. & Ichikawa, M. (2011a). Nanocontact heteroepitaxy of thin GaSb and AlGaSb films on Si substrates using ultrahigh-density nanodot seeds. *Nanotechnology* 22, 265301-1-7

Nakamura, Y.; Murayama, A. & Ichikawa, M. (2011b). Epitaxial Growth of High Quality Ge Films on Si(001) Substrates by Nanocontact Epitaxy. *Cryst. Growth Des.*, 11, 3301-3305

Niquet, Y. M.; Allan, G.; Delerue, C. & Lannoo M. (2000). Quantum confinement in germanium nanocrystals. *Appl. Phys. Lett.*, 77, 1182-1184

Ragan, R. & Atwater, H. A. (2000). Measurement of the direct energy gap of coherently strained Sn_xGe_{1-x}/Ge(001) heterostructures. *Appl. Phys. Lett.*, 77, 3418-3420.

Schmidt, O. G. & Eberl, K. (2000). Multiple layers of self-asssembled Ge/Si islands: Photoluminescence, strain fields, material interdiffusion, and island formation. *Phys. Rev. B*, 61, 13721-13729,

Shklyaev, A. A.; Shibata, M. & Ichikawa, M. (2000). High-density ultrasmall epitaxial Ge islands on Si(111) surfaces with a SiO_2 coverage. *Phys. Rev. B*, 62, 1540-1543

Shklyaev, A. A. & Ichikawa, M. (2002). Three-dimensional Si islands on Si(001) surfaces. *Phys. Rev. B*, 65, 045307-1-6

Shklyaev, A. A.; Nobuki, S.; Uchida, S.; Nakamura, Y. & Ichikawa, M. (2006). Photoluminescence of Ge/Si structures grown on oxidized Si surfaces. *Appl. Phys. Lett.*, 88 121919-1-3

Shklyaev, A. A.; Cho, S. P.; Nakamura, Y., Tanaka, N. & Ichikawa, M. (2007). Influence of growth and annealing conditions on photoluminescence of Ge/Si layers grown on oxidized Si surfaces. *J. Phys.: Condens. Matter*, 19, 136004-1-8

Matsudo, T.; Ohta T.; Yasuda, T.; Nishizawa M.; Miyata N.; Yamasaki S.; Shklyaev, A. A. & Ichikawa M. (2002). Observation of oscillating behavior in the reflectance difference spectra of oxidized Si(001) surfaces. *J. Appl. Phys.*, 91, 3637-3643

Wolkin, M. V.; Jorne, J.; Fauchet,P. M.; Allan, G. & Delerue, C. (1999). Electronic States and Luminescence in Porous Silicon Quantum Dots: The Role of Oxygen. *Phys. Rev. Lett.*, 82, 197-200

7

Self-Assembled InAs(N) Quantum Dots Grown by Molecular Beam Epitaxy on GaAs (100)*

Alvaro Pulzara-Mora[1], Juan Salvador Rojas-Ramírez[2], Victor Hugo Méndez García[3], Jorge A. Huerta-Ruelas[4], Julio Mendoza Alvarez[2] and Maximo López López[2]

[1]Laboratorio de Magnetismo y Materiales Avanzados, Universidad Nacional de Colombia Sede Manizales;
[2]Physics Department, Centro de Investigación y Estudios Avanzados del IPN, México D.F., México;
[3]Coordinación para la Innovación y Aplicación de la Ciencia y Tecnología, Universidad Autónoma de San Luis Potosí, San Luis Potosí, S.L.P, México
[4]Centro de Investigación en Ciencia Aplicada y Tecnología Avanzada, Instituto Politécnico Nacional Cerro Blanco 141 Colinas del Cimatario Querétaro, Querétaro México México

1. Introduction

The zero-dimensional nature of self-assembled quantum dots (SAQDs) is of great interest for high-performance optoelectronic devices such as semiconductor lasers [Kita et al., 2003]. Self-assembled Stranski–Krastanov (SK) growth mode is a promising technique to realize defect free and high SAQDS density. However, the size fluctuations of SAQDS observed in SK growth mode hinder the realization of devices superior performance.

The device performance strongly depends on the SAQDs parameters, such as the size, density, and uniformity, which can be controlled by performing precise growth. Most studies on SAQDs have been focused on InAs/GaAs systems and toward developments in optical fiber communication systems at wavelengths of 1.3 or 1.55 μm [Matsuura et al, 2004; A. Ueta et al, 2004]. However, it is difficult to obtain emissions in this wavelength region with InAs SAQDs, because of the strain effects inside them. InAs has a large lattice mismatch (~7%) to GaAs, and InAs SAQDs embedded inside structures are strongly compressed by their surrounding GaAs barriers. Recently, dilute nitride III–V compounds have been the subject of important research interest as they exhibit new properties that are potentially useful for narrow band gap devices. It has been found that replacing a small amount (< 5%) of the group V element by nitrogen in III–V compounds reduces the energy gap and changes the electronic structure, such as InAsN alloy due to its large bowing factor ($b \approx 16$) [J. Wu et al, 2002], thus offering new perspectives for band structure engineering in order to improve optoelectronic properties. Therefore, InAsN SAQDs are promising to obtain emissions at a wavelength of 1.55 μm and longer [L. Ivanova et al, 2008].

On the other hand, it is widely known that reflection high-energy electron diffraction (RHEED) is a powerful in-situ characterization technique, RHEED is quite sensitive to the crystallographic surface structures in real time. Many authors have used this technique to study the QD shape evolution during the GaAs capping layer growth. However, very few authors have studied the relation between diffraction pattern (chevrons) with the shape and size of quantum dots [H. Lee et al, 1998; T. Hanada et al, 2001b]. In order to minimize the effects of size and shape fluctuation, the growth parameters such as growth temperature and Arsenic overpressure must be optimized.

On the other hand, in a similar way as in GaAsN and GaPN alloys, the incorporation of Nitrogen (N) in InAs, yields a large bandgap bowing which produces a significant reduction of the bandgap energy. Therefore, InAsN SAQDs are promising to obtain emissions at a wavelength of 1.55 μm and longer.

In this work we have studied the effects of growth temperature and Arsenic overpressure on the growth mode of InAs(N) SAQDs on GaAs(100) substrates. The wavelength emission of the InAs(N) SAQDs was evaluated depending on growth conditions. We present an analysis of the asymmetric broadening of the Raman spectra using the confinment phonon model (CM).

2. Experimental

The SAQDs were grown on GaAs (100) substrates employing a Riber C21 MBE system equipped with solid sources for III-V materials, and standard reflection high-energy electron diffraction (RHEED) system. For structures requiring Nitrogen Atomic Nitrogen was produced by a radio frequency (RF) plasma source. Ultrahigh purity nitrogen was introduced into the plasma source using a mass flux controller and a leak valve. First, in the MBE chamber the substrates were heated up to 580 °C to remove the surface oxides under an As_4 flux. Then, in order to smooth out surface imperfections, a 500nm-GaAs buffer layer was grown at 580 °C. At the end of the buffer layer growth the surface exhibited a sharp (2x4) RHEED pattern. After the buffer layer growth, the temperature was gradually lowered to the desired value (between 480 and 510 °C) and InAs SAQDs deposition was initiated. At these substrate temperatures the GaAs surface reconstruction changed to a c(4x4). The InAs growth rate was 0,06 monolayer (ML) per second. The growth mode was in-situ monitored by RHEED. Different InAs SAQDs samples were grown by changing the growth temperature, but maintaining constant the fluxes of As_4 and In. For InAsN SAQDs growth the radio frequency plasma source was operated at 100 W with a Nitrogen flux of 0.1 sccm. The InAs SAQDs dimensions and density were determined by Atomic Force Microscopy (AFM) and High Resolution Scanning Electron Microscopy (HRSEM). Photoluminescence Spectroscopy (PL) measurements were carried out at 10 K by using a double monochromator, and an InGaAs cooled detector. The 488 nm line from an Argon laser was used as the excitation wavelength at a 120 mW excitation power.

The structural properties of the samples were studied by micro Raman spectroscopy at room temperature using a spot size of $2x2\mu m^2$, and employing the 632.8nm line of a He–Ne laser in a backscattering configuration.

3. Results and discussion

3.1.1 RHEED Pattern

The temporal evolution of the RHEED intensity variation of the transmission spot for the InAs SAQDs grown at substrate temperatures of (M1) 480 °C, (M2) 490 °C, and (M3) 510

°C, with identical Arsenic overpressure (3x10^{-7}Torr), are shown in Fig. 1. During the first stages of InAs deposition a small hump was observed in the RHEED intensity, then it decreased smoothly. Finally, the RHEED intensity rises until reaching a saturation value. The RHEED patterns changes were continuously monitored from the wetting layer formation to the nucleation of self-assembled InAs dots. The RHEED pattern taken along the [01-1] azimuth from the buffer layer just at the start of InAs growth (marked by a downwards dashed arrow in Fig. 1) showed a c(4x4) GaAs surface reconstruction. Streaky RHEED patterns were maintained during the wetting layer formation (time t_{2D} in Fig. 1) indicating a layer by layer growth (the end of t_{2D} is marked by the downwards arrows in Fig. 1). However, after the elapsed time t_{2D} of InAs deposition, the streaky RHEED patterns changed to spotty, and the so-called Chevron patterns were evidenced [J.W. Lee et al, 2004], indicating the formation of 3D-dimensional structures.

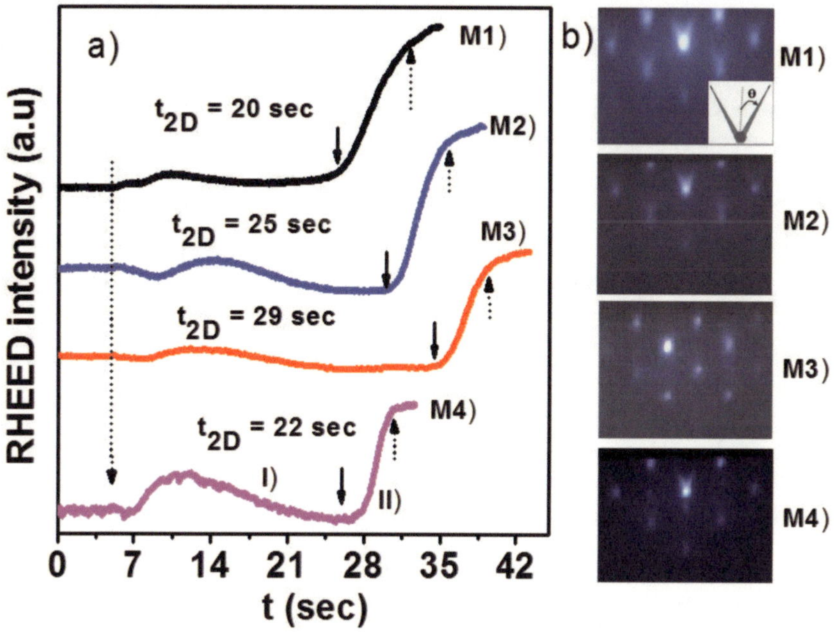

Fig. 1. (a) RHEED intensity of the transmission spot as a function of growth time (t) for InAs SAQDs grown at temperatures of: (M1) 480 °C, (M2) 490 °C, (M3) 510 °C, and (M4) 510 °C with Nitrogen. (b) RHEED patterns along the [01-1] after the growth of the InAs SAQDs. A constant Arsenic overpressure of 3x10^{-7} Torr was employed in all the growths.

The Chevron patterns consists of two well defined wings with an angle to each other of 2θ = 50 ° and 2θ = 56 ° for sample M1 and M2, respectively, as shown in the right column of Fig. 1. The fact that the Chevron´s angle increases by increasing the growth temperature, indicates that the slope of the facets that limit the SAQDS tends to increase with increasing growth temperature. The self-assembling process during the MBE growth is a non-equilibrium process, which is controlled by particular kinetics of diffusion and nucleation dependent on the growth conditions and the thermodynamics of the surface. The interplay of both, determines the ultimate morphology of the quantum dots, i.e., the facets that surround them. Nevertheless, in MBE growth usually the kinetic constrains seem to be predominant on the role of the intrinsic surface free energy. The change of the islands facets slope could be a result of an increase of the surface diffusion of In adatoms on the surface. For sample M3 the Chevron patterns are not well defined, rounded spots are observed similar to those present when the facets have the {163} crystal plane orientation.

In order to analyze the nitridation effects on InAs SAQDs formation, we grew an InAsN sample (M4) under identical conditions as for the sample M3 (T_g = 510 °C), but in a Nitrogen overpressure of $2x10^{-5}$ Torr. This temperature was employed in order to incorporate a low nitrogen content into the InAs SAQDs [A. Pulzara-Mora et al, 2007]. The RHEED intensity variation of the transmission spot for sample M4 (Fig. 1.) presented a similar behaviour as that for the InAs samples (M1-M3). However, the growth time for wetting layer formation (t_{2D}) decreases showing the strong effects of nitrogen incorporation into the SAQDs. Moreover, the Chevrons of RHEED patterns for sample M4 taken at the end of growth are better defined than those for sample M3, with a Chevron's angle of 2θ = 47.2°. The decrease of the Chevron's angle for sample M4 can be considered due to a decrease in the In surface diffusion caused by the additional Nitrogen overpressure. Note that a similar result has been observed when using a high Arsenic overpressure for the InAs QDs growth [T. Kudo et al, 2008]. These results suggest that the geometry and size of SAQDs can be changed by introducing Nitrogen during the InAs growth. We will discuss this point in the SEM and AFM images section.

Sample	Growth temperature (°C)	Angle (2θ) (degrees)	Facet planes
M1	480	49.84	(113)
M2	490	55.74	(113) tilted to (112)
M3	510	-	(136)
M4 (with Nitrogen)	510	47.2	(113) tilted to (114)

Table 1. Angle (2θ) of the Chevrons of RHHED patterns and the associated facet planes.

In table I we summarize the Chevron's angle (θ) in the RHEED patterns of SAQDs grown under different conditions, and the associated crystal planes of the facets. The corresponding (1 1 x) planes of the facets were obtained from the linear correlation between the tangent of the Chevron semi-angle θ/2 and the magnitude of the index x [A. Feltrin et al, 2007]. We found that

$$tan\left(\frac{\theta}{2}\right) = \frac{\sqrt{2}}{x} \quad (1)$$

3.1.2 Morphological characterization

Fig. 2 (I), shows a HRSEM image taken in cross-section view (x 300000) from sample M4. In this image is clearly observed the substrate (S), buffer layer (BL) and semi-spherical shaped SAQDs nucleated on the wetting layer. The InAs SAQDs have a height between 7 and 10 nm. The HRSEM image of the same sample in Fig. 4 (II) taken in plane view (x50000), shows that the SAQDs are not homogeneously distributed on the wetting layer, but they tend to nucleate on preferential regions. We think that those regions are surface steps on the GaAs (100) substrate [S. O. Cho et al, 2006], as illustrated in the sketch inserted in the micrograph.

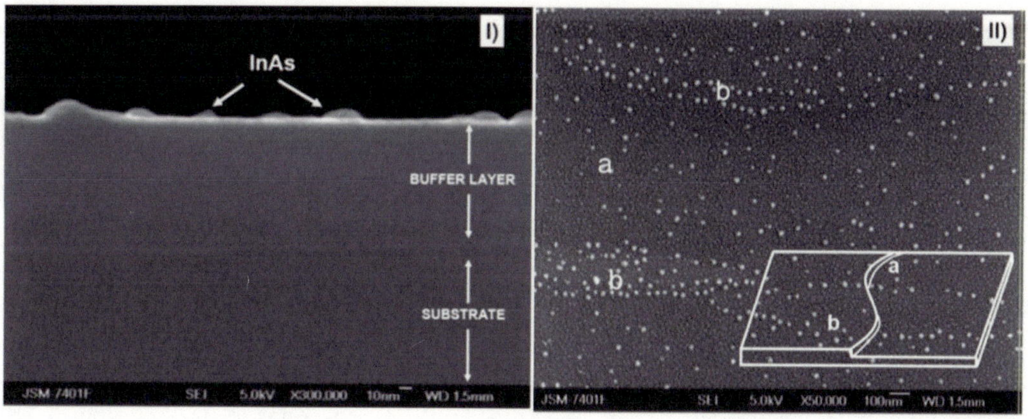

Fig. 2. (I) Cross-sectional HRSEM image (x300.000) of InAs SAQDs. (II) Plane view HRSEM image (x50.000) of InAs SAQDs. The inset shows schematically the preferential sites for SAQDs nucleation.

AFM images with a scanned area of 1×1 μm² of InAs dots on the (100) GaAs surface are shown in Figs. 3 (M1-M3). The dot densities obtained from a statistical analysis are: 6.84×10^{10} cm^{-2}, 3.4×10^{10} cm^{-2} and 1.87×10^{10} cm^{-2}, for M1, M2 and M3, respectively. We also observe that the average InAs dots size tends to increase with increasing the growth temperature, this could be due to an increase in the surface mobility of In adatoms [M. Sopanen et al, 2000; S. Kiravitaya et al, 2002]. On the other hand, the density of the InAsN SAQDs was 8.7×10^{9} cm^{-2}. By comparing the AFM images for samples M3 and M4, we observed that the density of InAsN SAQDs is one order of magnitude lower than that of sample M3 grown under identical Arsenic overpressure and substrate temperature. Moreover, we noted that the nitridation of InAs SAQDs promoted the formation of quantum dots of larger dimensions (≥10 nm).

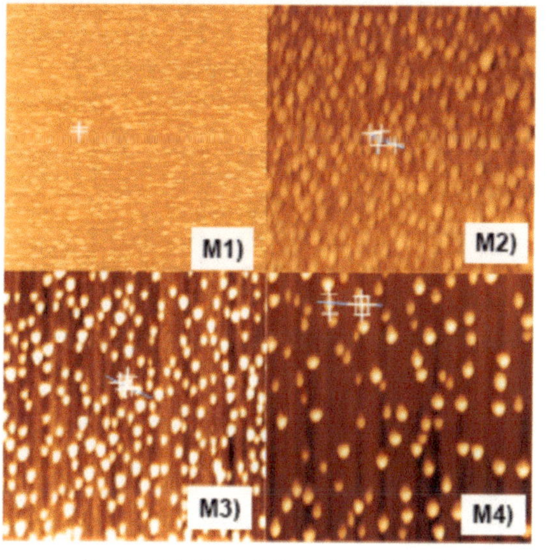

Fig. 3. 1x1 μm² AFM images of InAs SAQDs (M1, M2, and M3) and InAsN SAQDs (M4).

3.2 Optical characterization

Among the optical methods to characterize epitaxially grown quantum dots, photoluminescence and Raman spectroscopy are probably the most common. We used these techniques to find out the ground state transition of the InAs SAQDs, and analyzed the possible phonon scattering come from GaAs, InAs QDs, and surface phonon modes. All photoluminescence spectra were taken at 10 K. The Raman spectra were taken at room temperature by using a backscattering configuration.

3.2.1 Photoluminescence (PL)

In photoluminescence (PL), when a quantum dot is irradiated with light, a photon with energy $h\omega_{exc}$ is absorbed in the quantum dot or in the surrounding material, creating an electron-hole pair (excitation). Then, electron and hole is relaxed by moving to the energetically lowest state in the QD (relaxation).

The relaxation process for the electrons is through the creation of phonons, which have quantized energies. This effect known as phonon bottleneck has been reported for various authors [H. Benisty et al, 1981; R. Heitz et al, 2001]. Finally, the electron-hole pair recombines, emitting a photon with energy $h\omega_{rec}$ (recombination) which is detected as PL light. Fig. 4 shows the schematic representation of the PL process in a quantum dot. On the other hand, due to the incident light spot in the PL experimental setup is much larger than 1 μm^2, many QDs are excited simultaneously. Therefore, several series of peaks can be observed simultaneously, which result in one broad peak as shown by the experimental PL results. In a quantum dot the single particle energies of electrons or holes depend almost solely on the QD's structural properties like size, shape, composition and the surrounding material.

The single particle picture is, however, no longer sufficient if a QD is occupied by more than one charge carrier, because the Coulomb interaction between the confined particles alters the overall energy of the system. For bulk and quantum well structures this well-known effect is referred to as renormalization of the band gap. For fully confined systems like QDs such renormalization is also expected to play a role but to have only minor impact, since the strong confinement dominates the electronic properties of a QD.

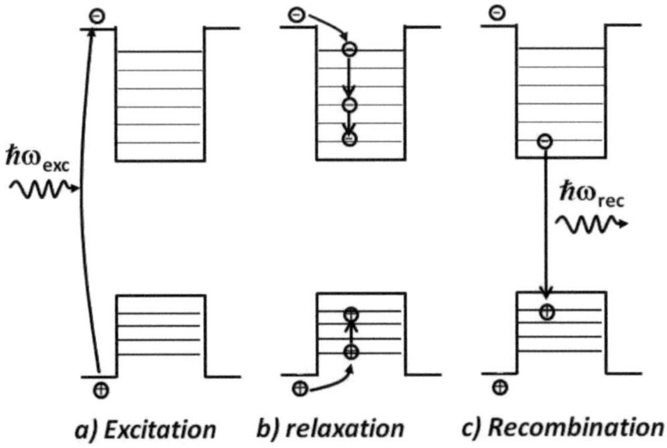

Fig. 4. Schematic representation of the photoluminescence processes in a Quantum dot

The PL band is the convolution of all the emission lines of every single QD which is excited; the lines have an homogenous broadening, due to optical recombination and scattering by phonons and impurities, and an inhomogenous one, related to differences in size, shape and composition of individual QDs. Therefore, the shape of the PL bands reflects the distribution of quantum confined energy levels in QDs [S. Franchi et al, 2003].

PL spectrum for the InAs and InAsN SAQDs are shown in Fig. 5(a). PL spectrum for the InAs and InAsN SAQDs are shown in Fig. 4(a). The PL spectra for the samples M1-M3 show a broad (100, 86, and 81 meV, respectively), symmetric main peak, coming from the statistical distribution of SAQDs. The position of the PL peak varies from the 1.23 μm to 1.52 μm depending on InAs SAQDs sizes. Great efforts have been made to obtain SAQDs emitting at longer wavelengths for a variety of applications. One possibility is to make an annealing on the samples in an ultra high vacuum atmosphere [O. Suekane et al, 2002], another possibility is to carry out a nitridation of InAs SAQDs after growth. However, the crystalline quality of the InAs dots is degraded, and the optical properties change drastically. Therefore, we preferred to carry out the nitridation during the InAs growth (sample M4). In this case the optical quality of the InAsN SAQDs didn't present considerable changes as can be seen in Fig. 5(a) (M4). The PL signal for sample (M4) has a long asymmetric tail at short wavelengths, probably due to incorporation of atoms or clusters of Nitrogen in the InAs SAQDs typical of III-V:N compounds [V. Sallet et al, 2005]. The PL peak of sample M4 is located at 1.55 μm, this red-shift could be caused by the following effects: 1) Nitrogen incorporation in the wetting layer, which decreases the compressive stress because $a_{InAs}>a_{InAsN}>a_{GaAs}$, where a_{InAs}, a_{InAsN}, and a_{GaAs} are the InAs, InAsN and GaAs lattice parameters, respectively. 2) Nitrogen incorporation in the SAQDs, as above mentioned the bandgap of III-V compounds is reduced by the incorporation of N. 3) The formation of larger SAQDs promoted by the incorporation of Nitrogen as observed in AFM images presented in the Fig. 4 (M4).

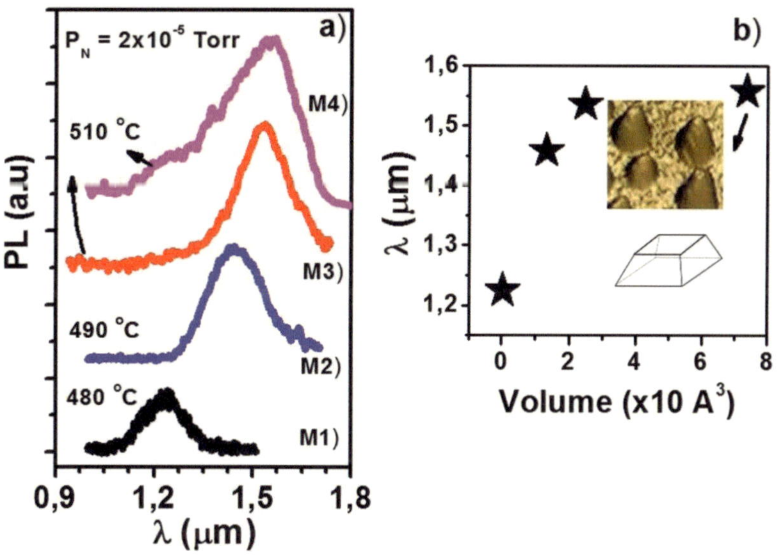

Fig. 5. (a) Low temperature PL spectra of InAs (M1, M2 and M3) and InAsN (M4) SAQDs. (b) Wavelength emission as a function of the SAQDs volume.

The PL peak position for the InAs(N) SAQDs is shown in Fig. 5(b) as a function of dots volume obtained from AFM measurements, and by using the geometry of a truncated pyramidal solid as it is shown in the inset of this figure. The wavelength of the PL emission increases for larger InAs SAQDs as a consequence of decreasing the quantum confinment.

3.2.2 Raman Spectroscopy (RS)

Raman spectroscopy on solids is nowadays widely known as a means of studying phonons, while it is also a very suitable method to study electronic collective excitations in semiconductors nanostructures, such as quantum wells or quantum dots. In the formation of SAQD, it is well known that strain plays an important role and Raman measurement is a strong tool to study strain. The mismatch of lattice constants gives rise to strain fields in QD nanostructures, which will affect their optical properties [L. B. Freund, 2000]. In-plane lattice mismatch parameters are defined as:

$$(\varepsilon_0)_{xx} = (\varepsilon_0)_{yy} = \frac{a_s - a_d}{a_d} \tag{1}$$

Where a_s and a_d are the lattice constants of the substrate (GaAs) and the dot (InAs) materials, respectively, which will be taken as 0.565 and 0.605 nm in this paper. Obviously, the lattice constant of the dot material (InAs) exceeds that of the substrate material (GaAs). Hence, it is expected that InAs will experience compression from GaAs, while GaAs will experience tension from InAs. Note that the lattice mismatch parameters defined in equation (1) do not describe completly strain fields in the quantum-dot island. In fact, lattice mismatches will induce elastic deformation in both the substrate and the island, to ensure the equilibrium of the corresponding stresses.

The evolution of the Raman FWHM and frequency shifts with increasing the QDs zise can also be described by a phenomenological model, the so-called "Phonon Confinement Model" (PCM). This model was originally developed to describe the evolution of the Raman spectra of disordered semiconductors [H. Richter et al, 1981; Tiong et al, 1984]. Later Campbell and Fauchet generalized this model to study the phonon confinement in nanostructures and thin films [I. H. Campbell, 1986]. PCM is widely used method in modeling Raman spectra of low dimensional systems such as quantum wells, quantum wires and quantum dots. Many experimental groups have used the PCM method to explain the observed asymmetry of the one-phonon bands appearing in their nanowire Raman spectra [K. W. adu et al, 2005; Giuseppe Faraci et al, 2006]. However, this method has not been explored in detail for analyzing Raman spectra of InAs SAQDs grown by MBE. We will discuss this point in detail in the next section.

In the phenomenological phonon confinement model (PCM), the Raman scattering intensity I(ω,d) for a spherical particle of diameter L, and Gaussian confinement function is given by [M. Grujic et al, 2009]

$$I(\omega) = I_0 \int_{BZ} |C(o,q)|^2 F[\omega_i(q), \Gamma] d^3q \tag{2}$$

where $|C(o,q)|^2 = e^{-\frac{q^2L^2}{2\beta^2}}$ is the Fourier coefficient of the confinement enveloped function $W(r) \propto e^{-\left[\frac{\alpha r}{L}\right]^2}$ for the zone centre optical phonon, q is the wave vector $\left[\frac{-\pi}{a} \leq q_{BZ} \leq \frac{\pi}{a}\right]$, a is the unit cell lattice parameter, and L is the average diameter. The confinement factor β initially was taken as a constant factor (Richter *et al*, β = 1, Campbell *et al.* β = 2π²). However, many authors consider that the confinement factor β must be a fitting parameter in Raman spectra, because it can be dependent on confinement boundary conditions in nanomaterials. $F[\omega_i(q),\Gamma]$ is the spectral line shape related to phonon disperse curve $\omega_i(q)$ with Γ natural line width (FWHM). Gaussian or Lorentzian are the most used lineshapes to fitting the experimental Raman data. However, due to the asymmetry and broadening of our experimental Raman spectra we chosen the so called Breit –Wigner-Fano (BWF) lineshape [Ugo Fano, et al, 1961], given as,

$$F[\omega_i(q),\Gamma] = \frac{Io\left[f\left[\frac{\Gamma}{2}\right]^2 + \omega_i(q) - \omega_{LO}\right]^2}{[\omega_{LO} - \omega_i(q)]^2 + \left[\frac{\Gamma}{2}\right]^2} \quad (3)$$

In which $1/f$ is the Fano parameter, and represents the asymmetry of the shape, while Io and Γ are fitting parameters of the intensity and broadening factor, respectively. The Fano line shapes evolve to Lorentzian line when f tends to zero. In the Eq (2), the infinitesimal elemental d^3q volume can be written as q^2dq if a spherical symmetry is considered.

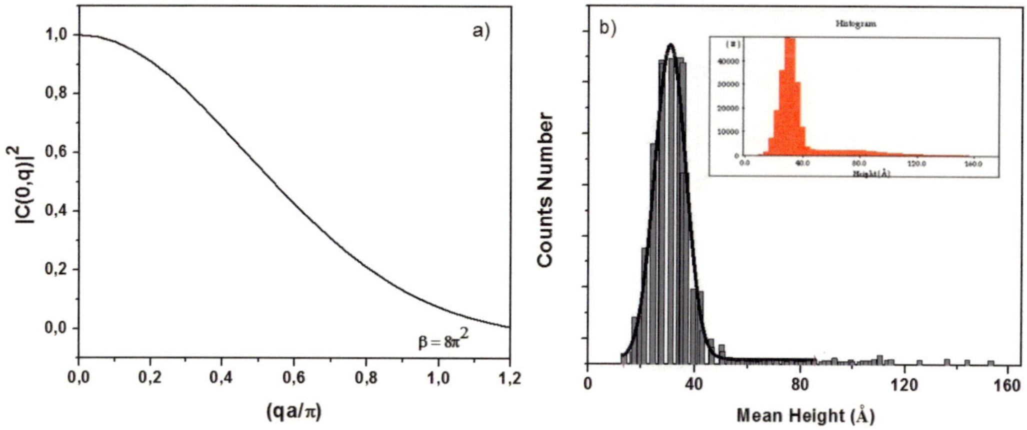

Fig. 6. (a) Squares of the Fourier transform $|Q(0,q)|^2$ used in the confinement phonon model (CM), β = 8π². (b) Diameter distribution of the InAs Qdots (M4-sample) with mean height L = 35 Å. The solid curve is a log-normal fit to the distribution. The inset in (b) is the histogram of the sample M4.

GaAs has two atoms in the unit cell and, therefore, six phonon branches. Three acoustic and three optical. From the three optical branches, one gives rise to an infra-red active mode at the Γ point, while the two other branches are degenerate at the Γ point and Raman active. Therefore, zone center (q = 0) phonons would generate a one peak Raman spectra. However, the electronic structure of GaAs generates special electron-phonon induced resonance conditions with non-zone center modes (q ≠ 0), and the previous rule is not valid. In this case, for the LO phonon frequency dispersion, we can use the diatomic linear-chain model, which is expressed as,

$$\omega(q) = C\sqrt{\frac{M_1+M_2\sqrt{{M_1}^2+{M_2}^2+2{M_1}^2{M_2}^2\cos(qa)}}{M_1M_2}} \quad (4)$$

where M1 and M2 are the atomic masses of Ga and As, respectively and C is the fitting parameter. (a_{GaAs} = 0.5466 nm).

Fig. 7 shows the Raman spectra taken at room temperature by using the 632 nm excitation laser line. The two peaks at 265 and 290 cm^{-1} are due to scattering from the GaAs TO and LO phonons [Tomoyuki Sekine et al, 2002], respectively. Since Raman measurement is performed in backscattering configuration from (001) growth surface, TO mode should be forbidden. The low intensity of TO mode is due to the departure from the perfect backscattering geometry. It is observed that the line center of these peaks remains practically the same in all the samples. However, by comparing all spectra one can see that the peak position of the TO and the LO peaks are up-shifted about 4.7 cm^{-1} compared with the bulk (001) GaAs (TO = 260 cm^{-1}, LO = 290 cm^{-1}). The compressive stress acting on the QDs can produce produce a blue-shift in the LO-phonon frequency. Presumably, the red-shift induced by phonon confinement compensates for the blue-shift due to the surface pressure effects entirely as has been observed in some semiconductors materials [Campbell I.H. & Fauchet P.M., 1986], embedded in a glass matrix.

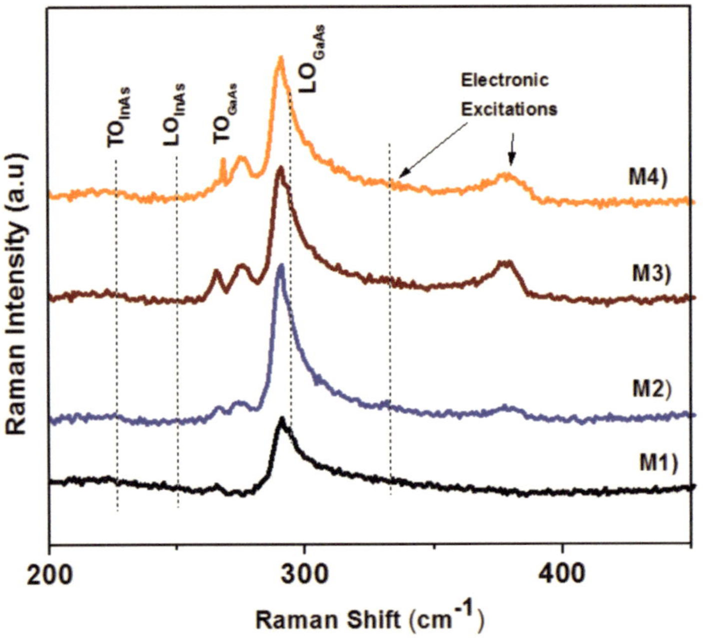

Fig. 7. Raman spectra taken at room temperature of the InAs QDots

On the other hand, it is also observed that the GaAs LO mode is asymmetrically broadened compared to the bulk, as shown the Fig. 6). The asymmetry observed in the one phonon LO Raman has been tentatively assigned to several factors such as particle size distribution, random variations in the bond wave number, and particle shape irregularity [AkhileshK et al, 2007]. Quantum confinement effect can contribute to the asymmetrical broadening observed in the one-phonon LO Raman (Fig 7. and Fig. 8.).

Experimental results show that as the QD volume increases (samples M1 to M4), there is a change in the line shape and full-width at half-maxima (FWHM) of the LO-phonon peaks.

The Raman peak of M4 sample with quantum dots of 73Å^3 volume, has an FWHM of 8.2 cm^{-1}. As the QD volume decreases from 2.4 Å^3 to 0.26 Å^3 the FWHM increases to ≈ 9.8 cm^{-1}. The peak profile is almost symmetric in the case of sample M4, whereas a small asymmetry in the line shape becomes noticeable for the samples in the strongly confined regimen. This observation is more evident for sample M1 with the smallest SAQDs volume

In order to analyze the asymmetry and broadening of the Raman spectra, we carry out the integration of the equation (2) using the dispersion relation of linear diatomic chain gives by the eq. (4). To assess the quantum confinement effect in the Raman spectra by reducing the volume of quantum dots, the measured average diameter, was also included in the Richter line shape equation (2). Then a second integration with respect to L was carried out. F(L) is the log-normal distribution function obtained by fitting the histogram from AFM measurements.

$$F(L) = \frac{1}{\sigma} e^{-\left[\frac{\ln(L)-\ln(\bar{L})}{2\gamma^2}\right]} \quad (5)$$

Where $\bar{L}$ *and* σ are the mean diameter and width parameter, respectively. Eq. (5) allows introduce the diameter distribution of the QDots analytically into the Richter Raman line shape equation up (Eq. 2). Finally, the Ricther line shape then become

$$I(\omega, L) = I_0 \int_{BZ} \mathrm{F(L)I}(\omega)\mathrm{dL} \quad (6)$$

In order to analyze the broadening and asymmetry of the experimental Raman data, we used the model of three-dimensional phonon confinement. Raman spectra of InAs QDs were calculated using the equations (6), (4), and (3) and confinment function. In the Fig. 8, we show the result of our fitting for the experimental 290 cm^{-1} Raman band from samples M1-M4. In all fitting we only varied the intensity Io, the confinement parameter (σ) and 1/f Fano-parameter. The other parameters used in this fitting are show in the table 2. The spectra show a blue-shift of the LO GaAs mode, and an asymmetric broadening with decreasing QDs volume. We found that the phonon confinement model qualitatively explains the asymmetric lineshape and the increase in the FWHM with a decrease in the QDs volume. For the sample M1 with lower volume confinement, the spectrum calculated with the CM model differs considerably from the experimental data. By increasing the volume of confinement, (M2-M4 samples) the spectrum best fits the experimental data. Several authors report that the CM model explains very well the Raman spectra for nanocrystals (e.g. In or Ge QDs) with a diameter higher than 4 nm. This difference has been interpreted theoretically to be originated from the different crystallographic natures of their Raman spectra: crystalline for nonpolar and amorphous for polar nanosemiconductors. On the other hand, the wave function used in the quantum confinement model (CM) is of a single electron, and a single vibrational mode. However, in a finite crystal the vibrational modes depend on different values of the wave vector k_m [$k_m = \frac{m\pi}{Na} = \frac{m\pi}{L}$, m= 2,4,6,.... Max(a/D), where D is the diameter of crystallite], as a consequence of the boundary condition. In this case, an additional term is needed in the Eq. (6), but it is not the purpose of this work.

Table 2, shows the FWHM of the peak centered at 290 cm^{-1} (Fig. 7.); the mean height $\bar{L}$ $(\dot{A})$ and the FWHM obtained from fitting the AFM histogram with a log-normal distribution as shown in the inset of the Fig. 6. b.

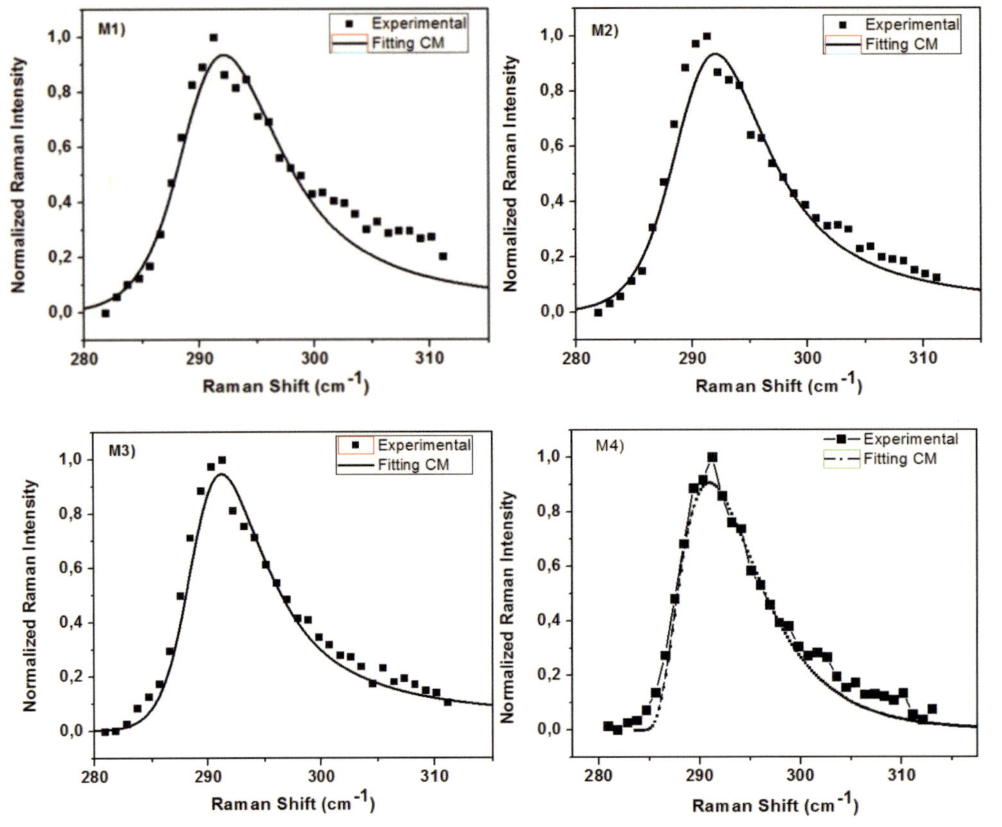

Fig. 8. Fitting on Raman spectra to the Confinement Model (CM). The square points are the experimental data and the continuous line is the CM by using the Eq(6).

Samples	FWHM Raman spectra (cm^{-1})	Mean height $\bar{L}$ ($\dot{A}$)	FWHM QDots distribution $\gamma(\dot{A})$
M1	9.8	17	35
M2	9.1	25	30
M3	9.0	30	24
M4	8.2	35	14

Table 2. Experimental data used to fit the Raman spectra

In the low frequency region of the LO phonon peak, we have observed a set of additional features in Raman spectra which can be associated to the SADQs (labeled by arrow, in Fig. 7., as known as *surface optical phonons* (*SO*). While phonons are collective lattice vibration modes, surface phonons are particular modes associated with surfaces; they are an artefact of periodicity, asymmetry, and the termination of bulk crystal structure associated with the surface layer of a solid [P. Nandakumara et al, 20001]. The SO-mode can be observed in the Raman scattering processes due to: a) impurity or interface imperfections, b) valence band mixing arising from the degeneracy at the Γ point in the Brillouin zone of zinc-blende semiconductors, and c) non-spherical geometry of the QDs. The study of surface phonons provides valuable insight into the surface structure and other specific properties from the surface region, which often differ from bulk. Owing to the small size in QD's, the surface-to-volume ratio contribution to the Raman spectrum is much higher than in bulk crystals. In nanocrystals such as InAs QDots these modes can appear when the nanostructures are surrounded with a material of different dielectric constant ω(ε), e.g air (ε = 1). In large crystals, phonons propagate to infinity and the 1st order Raman spectrum only consists of q=0 phonon modes. When crystalline perfection is destroyed due to lattice disorder and defects, symmetry forbidden modes (like SO phonon modes) are activated and become stronger with increasing defect density. SO-phonon modes are localized to the interfase, and can propagate along the interface. They are solutions to Maxwell's equations with appropriate boundary conditions [B. E. Sernelius, 2001]. Many authors have studied theoretically the SO-modes from different geometries (cylindr ical and spheroids shape) in nanoparticles, nanowires and QDs, by using the approximation of a dielectric continuum and modeling of valence-force fields to explain the connection between SO phonon modes and the geometrical shape of a semiconductor QD [][Watt M et al, 1990; -37, Gupta R et al, 2203 -38].

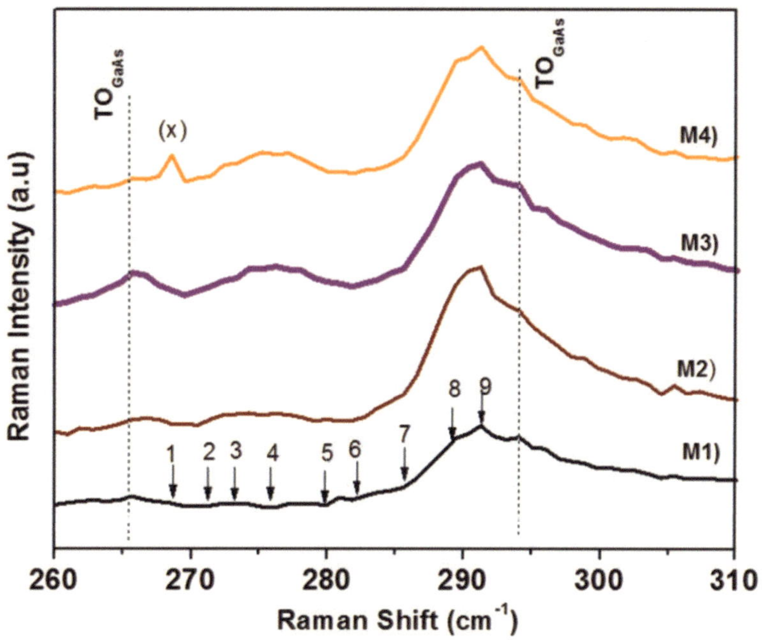

Fig. 9. Enlargement of the Raman spectra of the InAs SAQDs in the surface optical (SO) region. The arrows indicate small features related to the Rabi-type oscillations between two exciton levels. (x) in the sample M5 denote a phonon mode related to InAsN.

In our samples we observed that the integrated Raman intensity, and the position of the SO-mode for the samples M3 - M4 is clearly dependent of the volume and facets formed on SADQs. For sample M1 with quantum dots of smaller volume, and facets oriented along (113), a set of surface phonon modes indicated by a down-arrow (labeled from 1 to 9) are clearly identifiable. By increasing the confinement volume of quantum dots, M2 - M4 samples, (SO) phonon modes evolve in a continuum of states due to lowering of phonon quantum confinement. We believe that the splitting of the phonon modes in this region probably is related to combined effect of SO-surface and exciton oscillating modes. Vasilevskiy *et al* [M. I. Vasilevskiy 2005], using a non-perturbative approach to calculate the multi-phonon resonant Raman scattering probability in semiconductor QDs, show that when inter-level coupling is strong, in the first order Raman spectrum appear features similar to those obtained in the electron spectral function, effect known as Rabi splitting [S. Hameau et al, 1999]. They correspond to the exciton oscillating between its two states and (many times) emitting and absorbing a phonon. The electron–phonon interaction is important in semiconductors and, especially, in semiconductor nanostructures since it determines hot carrier relaxation, influences light absorption and emission processes and is responsible for Raman scattering. The intensity of this interaction is enhanced in quantum dots (QD's) and, owing to the discrete nature of the electron energy spectrum, it leads to multi-phonon processes and formation of a polaron.

By comparing the position of SO-mode of the samples M2 and M3 (surrounded by air, $\varepsilon = 1$) with the nitrogenated M4- sample position (surrounded with N, $\varepsilon_N = 1.454$), we find an up-shift of 1.0 cm^{-1}. It is show a dependence of SO-mode with a systematic increase in the dielectric function that is surrounding the QDs. Further SO-modes, an additional peak (268.5 cm^{-1}, labelled (x)) from LO-InAsN phonon mode was only observed in the nitridated -M4 sample. In addition of the confined optic phonons, the presence of surface phonons in the Raman spectra of nanostructured of materials has been reported in a number of systems [Ingale A. et al, 1998; Alim et al, 2005]. Recently Ladanov et al [M. Yu et al, 2005] reported theoretical calculation of surface modes (SO) on spherical shape InAs QDs surrounded by a AlAs layer and air using the approximation of a dielectric continuum. They found that the frequencies of interface phonons (SO) obtained within this model lie in the spectral range between the frequencies of TO and LO phonons, and depend on the shape of the quantum dots.

On the other hand, in the region from 310 cm^{-1} to 500 cm^{-1} (Fig. 7.) we observed a set of peaks associated to electronic transitions. The Raman signal from electronic excitations in quantum dots is expected to be very small unless we can take advantage of some resonance mechanism. In order to enhance the resonance Raman signal, we used a laser with a energy (1.9 eV) very close to the $E_0+\Delta$ gap for InAs quantum dots (1.7 eV measured from PL resonant experiments by W. Dasilva *et. al.* [W. da Silva et al, 1997]). The energy difference between the electronic states which can be measured in Raman and FIR absorption spectroscopy is on the order of 50 meV. The first observation of electronic excitations in self-assembled QDs was reported by L. Chu and coworkers [L. Chu et al, 1997; L. Chu et al, 2000]. The sample studied contained 15 layers of InGaAs QDs, each layer having a n-type GaAs doping layer in its vicinity. The number of electron confined into each QD was estimated to be $N_e = 6$. By using an excitation laser at an energy of 1.71 eV very close to $E_0+\Delta$ gap, they found a peak at about 50 meV (403.2 cm^{-1}) in the depolarized Raman spectra, which was interpreted as a spin density excitation (SDEx) with a transition between the p and d level (with general quantum number N=2 and N=3) in the quantum dots. In our experimental resonant Raman spectra (Fig 6) we observed a similar peak located around of 46 meV (377.1 cm^{-1}) for the samples M3-M5 (with N) in well agreement with previous report. However, for the samples M1 and M2 with smaller QDs volume a set of Raman signal coming from electronic transitions was not observed, due to small number of electrons confined in quantum dots by reducing the confinement volume.

4. Conclusions

Self-assembled InAs(N) quantum dots were grown on GaAs (001) substrates by molecular beam epitaxy. We studied the variation in geometry and density of SAQDs for different growth temperatures. The average InAs dots size tends to increase whilst their density tends to decrease when the growth temperature is increased, possibly due to an increase in the surface mobility of In adatoms. We observed that the nitridation of InAs SAQDs starting from the wetting layer formation promoted the formation of quantum dots of larger dimensions.

5. Acknowledgments

This work was partially supported by CONACyT-Mexico, ICTDF-Mexico; DIMA and COLCIENCIAS-Colombia. The authors thank the technical assistance of R. Fragoso, A. Guillen, Z. Rivera, and E. Gomez

6. References

A. Pulzara-Mora, E. Cruz-Hernández, J. Rojas-Ramirez, R. Contreras-Guerrero, M. Meléndez-Lira, C. Falcony-Guajardo, M.A. Aguilar-Frutis, and M. López-López. (2007). Study of optical properties of GaAsN layers prepared by molecular beam epitaxy. *Journal of Crystal Growth,* Vol 301-302, (April 2007), pp. 565-569. doi:10.1016/j.jcrysgro.2006.11.241

A. Ueta, S. Gozu, K. Akahane, N. Yamamoto, and N. Ohtani. (2005). Growth of InAsSb Quantum Dots on GaAs Substrates Using Periodic Supply Epitaxy. *Japanese Journal of Applied Physics*, Vol. 44, (May 2005), pp. L696-L698, ISSN 0021-4922.

Adu K.W., Gutierrez H.R., Kim U.J., Sumanasekera G.U. & Eklund P.C. (2005). Confined phonons in Si nanowires. *Nano Lett.* 5, 409 , ISSN: 1530-6984

Alim K, Fonoberov V, Balandin. (2005). Origin of the optical phonon frequency shifts in ZnO quantum dots, *Appl. Phys. Lett.*; 86: (January 2005), p.p. 053103-3. ISSN 0003-6951.

B. E. Sernelius. (2001). *Surface Modes in Physics*, 1st edn. (Wiley-VCH, New York), p. 350. ISBN 3-527-40313-2

Benisty, H.; Sotomayor-Torres, C. M. & Weisbuch, C. (1991). Intrinsic mechanism for the poor luminescence properties of quantum-box systems, *Physical Review B*, Vol. (November 1991)) 44, p.p 10945–10948. ISSN 10945-10948.

Fauchet P.M. & Campbell I.H. (1988). Raman-spectroscopy of low-dimensional semiconductors. *Critical reviews in solid state and materials sciences* 14, p,p S79-101, ISSN: 0161-1593.

Gupta R., Xiong Q., Mahan G.D. & Eklund P.C., (2003). Surface optical phonons in gallium phosphide nanowires. *Nano Lett.*, 3, (Dec 2003), p.p 1745-50, ISSN: 1530-6984.

H. Lee, R. Lowe-Webb, W. Yang, and P. C. Sercel. (1998). Determination of the shape of self-organized InAs/GaAs quantum dots by reflection high energy electron diffraction. *Applied Physics Letters*, Vol. 72, No. 7, (February1998), pp. 812-814, ISSN 0003-6951.

Ingale A, Rustagi KC. (1998). Structural and particulate to bulk phase transformation of CdS film on annealing: A Raman spectroscopy study, *Phys. Rev., B*; 58: (October 2009), p.p. 7197-204.. ISSN 0021-8979.

J. Wu, W. Walukiewicz, K. M. Yu, J. W. Ager, III, E. E. Haller, Y. G. Hong, H. P. Xin, and C. W. Tu. (2002). Band anticrossing in GaP1-xNx alloys. *Physical Review B*, Vol. 65, No. 24, (June 2002), pp. 241303-241306, ISSN 1098-0121.

J.W. Lee, D. Schuh, M. Bichler, and G. Abstreiter. (2004). Advanced study of various characteristics found in RHEED patterns during the growth of InAs quantum dots on GaAs (0 0 1) substrate by molecular beam epitaxy. *Applied Surface Science*, Vol. 228, No. 1-4, (April 2004), pp. 306-312. ISSN 0169-4332.

L. B. Freund. Int. J. Influence of strain on functional characteristics of nanoelectronic devices. (2001). *Solids Struct.* 37 (august 2001) p.p 1925-1935. doi:10.1016/S0022-5096(01)00039-4.

L. Chu, A. Zrenner, M. Bichler, G. Böhm, G. Abstreiter. (2000). Raman spectroscopy of In(Ga)As/GaAs quantum dot. *Appl. Phys. Lett.* 77 (24), (october 2000), p.p 3944-3946. ISSN 0003-6951.

L. Ivanova, H. Eisele, A. Lenz, R. Timm, M. Dähne, O. Schumann, L. Geelhaar, and H. Riechert. (2008). Nitrogen-induced intermixing of InAsN quantum dots with the GaAs matrix. 92 (2008) 203101. *Applied Physics Letters*, Vol. 92, No. 20, (May 2008), pp. 203101-203103, ISSN 0003-6951.

M. Grujic - Brojcin, M.J. S. cepanovic, Z. D. Dohcevic-Mitrovic and Z.V. Popovic. (2009). *Acta physica polonica* A, Vol. 116, p.p 51-54. Proceedings of the Symposium A of the European Materials Research, Warsaw, September 2008.

M. I. Vasilevskiy and R. P. Miranda. (2005). Is polaron effect important for resonant Raman scattering in self-assembled quantum dots?. *phys. stat. sol. (c)* 2, No. 2, (February 2005), p.p 862–866. DOI 10.1002/pssc.200460352.

M. Sopanen, H. P Xin, and C. W. Tu. (2000). Self-assembled GaInNAs quantum dots for 1.3 and 1.55 μm emission on GaAs. *Applied Physics Letters*, Vo. 76, No. 8, (February 2000), pp. 994-996, ISSN 0003-6951.

M. Watt, C. M. Sotomayor Torres, H. E. G. Arnot and S P Beaumont. (1990). Surface Phonon Modes in GaAs Cylinders, *Semicond Sci Technol* 5, issue 4 (April 1990), pp. 285-290. ISSN 0268-1242.

M. Yu. Ladanov, A. G. Milekhin, A. I. Toropov, A. K. Bakarov, A. K. Gutakovskii, D. A. Tanne, S. Schultze, and D. R. T. Zahn. (2005). *Journal of Experimental and Theoretical Physics*, Vol. 101, No. 3, (april 2005) pp. 554–561. Origin: CROSSREF; ADS. DOI: 10.1134/1.2103225.

Osamu Suekane, Shigehiko Hasegawa, Toshiko Okui, Masahiro Takata and Hisao Nakashima. (2002). Growth Temperature Dependence of InAs Islands Grown on GaAs (001) Substrates. *Jpn. J. Appl. Phys.* 41 (November 2001) pp. 1022-1025. 10.1143/JJAP.41.1022.

P. Nandakumara, C. Vijayana, M. Rajalakshmib, Akhilesh, K. Arorab. (2001). Raman spectra of CdS nanocrystals in Nafion: longitudinal optical and confined acoustic phonon modes. *Physica E* 11, ISSUE No 4. 377–383. (November 2001), PII: S 1386-9477(01)00157-6.

R. Heitz, H. Born, F. Guffarth, O. Stier, A. Schliwa, A. Hoffmann, and D. Bimberg. 2001. Existence of a phonon bottleneck for excitons in quantum dots. *Physical review b*, volume 64, (November 2001), p.p. 241305-1 - 241305-4, ISSN:1098-0121.

Richter H., Wang Z.P. & Ley L. (1981). The one phonon Raman-spectrum in microcrystalline silicon. *Solid State. Commun.*, 39, P.P 625-629, ISSN: 0038-1098.

S. Franchi, G. Trevisi, L. Seravalli, P. Frigeri. (2003). Quantum dot nanostructures and molecular beam epitaxy. *Progress in Crystal Growth and Characterization of Materials* 47 (April 2005) p.p. 166-195. doi:10.1016/j.pcrysgrow.2005.01.002. Available online 7 April 2005.

S. Hameau et al. (1999).. Strong Electron-Phonon Coupling Regime in Quantum Dots: Evidence for Everlasting Resonant Polarons, *Phys. Rev. Lett.* 83, (February 2995), p.p 4152 - 4155. ISSN 1089-7550.

S. Kiravitaya, Y. Nakamura, O. G. Schmidt. 2002. Photoluminescence linewidth narrowing of InAs/GaAs self-assembled quantum dots. *Physica E* 13, Volume 13, Issues 2-4 (January 2002), p.p 224-228 . ISSN 1386-9477.

S. O. Cho, Zh. M. Wang, and G. J. Salamo. (2005). Evolution of elongated (In,Ga)As–GaAs(100) islands with low indium content. *Applied Physics Letters*, Vol. 86, No. 11, (March 2005), pp. 113106-113108, ISSN 0003-6951(20050314)86:11;1-T.

T. Hanada, B. H. Koo, H. Totsuka, and T. Yao. (2011). Anisotropic shape of self-assembled InAs quantum dots: Refraction effect on spot shape of reflection high-energy electron diffraction. *Physical Review B*, Vol. 64, No. 16, (October 2001), pp. 165307-165312, ISSN 1098-0121.

T. Kita, Y. Masuda, T. Mori and O. Wada. (2003). Long-wavelength emission from nitridized InAs quantum dots. *Applied Physics Letters*, Vol. 83, No. 20, (November 2003), pp. 4152-4153, ISSN 0003-6951(20031117)83:20;1-Y.

T. Kudo, T. Inoue, T. Kita, and O.Wada. (2008). Real time analysis of self-assembled InAs/GaAs quantum dot growth by probing reflection high-energy electron diffraction chevron image. *Journal of Applied Physics*, Vol. 104, No. 7, (October 2008), pp. 074305-074309. doi:10.1063/1.2987469.

T. Matsuura, T. Miyamoto, T. Kageyama, M. Ohta, Y. Matsui, T. Furuhata, and F. Koyama. (2004). Elongation of Emission Wavelength of GaInAsSb-Covered (Ga)InAs Quantum Dots Grown by Molecular Beam Epitaxy. *Japanese Journal of Applied Physics*, Vol. 43, (January 2004), pp. L82-L84, ISSN 0021-4922.

Tomoyuki Sekine, Kunimitsu Uchinokura, Etsuyuki Matsuura. 2002. Two-phonon Raman scattering in GaAs. *Journal of Physics and Chemistry of Solids*. Volume 38, Issue 9 (September 2002),p.p 1091-1096. doi:10.1016/0022-3697(77)90216-5. Available online 23 September 2002

Ugo Fano (1961). Effects of Configuration Interaction on Intensities and Phase Shifts. *Phys. Rev.* 124, pp. 1866–1878. doi:10.1103/PhysRev.124.1866.

V. Sallet, L. Largeau, O. Mauguin, L. Travers, and J. C. Harmand. (2005). MBE growth of InAsN on (100) InAs substrates. *Phys. Stat. Sol.* (b) 242 (march 2005) R43 - R45. ISSN: 1521-3951.

W. da Silva, Yu. A. Pusep, J. C. Galzerani, D. I. Lubyshev, P. P. González-Borrero, and P. Basmaji(1997). Photoluminescence study of spin - orbit-split bound electron states in self-assembled InAs and $In_{0.5}Ga_{0.5}As$ quantum dots. J. Phys.: Condens. Matter. 9, Issue 1 (January 1997), p.p L13-L17. doi:10.1088/0953-8984/20/35/354007.

W. Walukiewicz. (2004). Narrow band gap groput III-nitride alloys. *Physica E* 20, (March 2005), p.p 300 – 307, ISSN: 1386-9477.

8

Hydrothermal Routes for the Synthesis of CdSe Core Quantum Dots

Raphaël Schneider[1] and Lavinia Balan[2]
[1]*Laboratoire Réactions et Génie des Procédés (LRGP) UPR 3349, Nancy-University, CNRS, Nancy Cedex*
[2]*Institut de Science des Matériaux de Mulhouse (IS2M)CNRS, LRC 7228, Mulhouse France*

1. Introduction

Semiconductor nanocrystals, also called quantum dots (QDs), are fluorescent inorganic particles with typical diameters ranging from 1 to 10 nm. Due to their quantum confinement, QDs show unique and fascinating optical properties that are advantageous in the fields of photovoltaic devices, light-emitting diodes, biological imaging, biodiagnostics,… (Costa-Fernandez et al., 2006; Lim et al. 2007; Medintz et al., 2005; Robel et al., 2006). CdSe QDs are probably the most extensively investigated colloidal II-VI semiconductor nanoparticles because (i) their band gap can be tuned across the visible spectrum by variation of their diameters, and (ii) of the advances made in their preparation.

Presently, CdSe QDs can be prepared in a high quality by using the organometallic synthesis (Qu et al., 2002; Foos et al., 2006). The hot injection technique used in that synthesis procedure produces a "burst nucleation" event, which is a crucial factor for the narrow size distribution of the nanoparticles. A reaction temperature of ca. 300°C is generally required to decompose the precursors used for the production of CdSe QDs. The hot-injection process involves complex manipulation that may limit its application in the scaled-up production and control of particle size, due to the difficulty in controlling the reaction temperature. In addition, it is not easy to obtain the desired products with the expected fluorescence due to the rapid growth rate of nanocrystals, which leads to a quick temporal evolution of the optical properties.

The choice of the ligands used to stabilize the nanoparticles is also crucial to obtain samples with a narrow size distribution and to control the shape of the nanocrystals. Fatty acids, fatty amines, phosphines, and phosphonic acids turned out to be suitable ligands allowing controlling the structural and electronic properties of QDs (Foos et al., 2006; Owen et al., 2008; Peng et al., 2001). Preparation methods for the synthesis of high quality and nearly monodisperse CdSe QDs have typically utilized tri-n-octylphosphine oxide (TOPO) and tri-n-octylphosphine (TOP) as these compounds provide the most controlled growth conditions. The hydrophobicity of TOPO and other ligands on the CdSe surface renders these QDs only dispersible in organic solvents of low polarity. However and especially for bio-applications, the ligands covering the surface of the as-synthesized CdSe nanoparticles are not suited for their further application and a modification of the surface properties is required. The most common method to render nanoparticles water-dispersible is to modify their surface with thiol-containing bi-functional ligands, such as thioacids and thioamines (Aldeek et al., 2011). This functionalization can lead to increases in hydrodynamic radii, as well as to instability when ligand exchange is not complete. Moreover, a loss of photoluminescence quantum yields (PL QYs) is often observed. QDs can also be encapsulated by a shell of material such as polymer, micelle or bead that makes them more dispersible in aqueous media. Such encapsulation significantly increases the diameter of the QD-based material, which may not be desirable in some applications such as biosensors and live cell imaging.

Because of this, the synthesis of QDs in aqueous solution is still pursued in hopes of providing a material that is easily fabricated and functionalized. During many years, the synthesis of CdSe in aqueous media has been investigated with limited success due to the low quantum yields and poor crystallinity of the nanoparticles produced. Moreover, the fluorescence of the CdSe QDs obtained by this approach cannot be controlled over a wide range. Direct synthesis of CdSe QDs in water is however a promising alternative route to organometallic reactions and facilitates the use of the dots in biological systems. Hydrothermal synthesis offers also the following advantages: (1) lower reaction temperatures with comparable PL QY; (2) does not use toxic and expensive organometallic reagents; (3) surface functionalization during synthesis without further treatment; (4) comparatively smaller sizes (3-8 nm) than those obtained after encapsulation of hydrophobic QDs with amphiphilic lipids or polymers (generally > 20 nm); (5) more reproducible.

Thus, it is still a challenging task to develop a method for preparing CdSe QDs with the desired quality under mild and environmentally friendly conditions using a low-cost and simple method.

In this review, we summarize the aqueous solution-based syntheses of CdSe nanocrystals developed in recent years and the applications of these nanocrystals.

2. CdSe photoluminescence

CdSe QDs exhibit size-dependent absorption and photoluminescence spectra which result from three-dimensional carrier confinement. The size of CdSe QDs can be used to tune the optical gap across a major portion of the visible spectrum. For example, the optical gap can be tuned from deep red (1.7 eV) to green (2.4 eV) by reducing the dot diameter from 20 to 2 nm. (Alivisatos, 1996).

PL studies have also demonstrated the important role of states in the nanocrystal energy band gap associated with the surface and/or defects.

The PL spectrum of CdSe QDs may be composed of two bands:

- the high energy band (band-edge emission) can be explained by different recombination mechanisms, such as recombination of the delocalized electron-hole (e-h) pairs or recombination through localized states, possibly of surface origin
- the low energy band in the near-IR-red spectral range originates from donor-acceptor recombination involving deep defect states associated to vacancies (Klimov et al., 1996) (Figure 1).

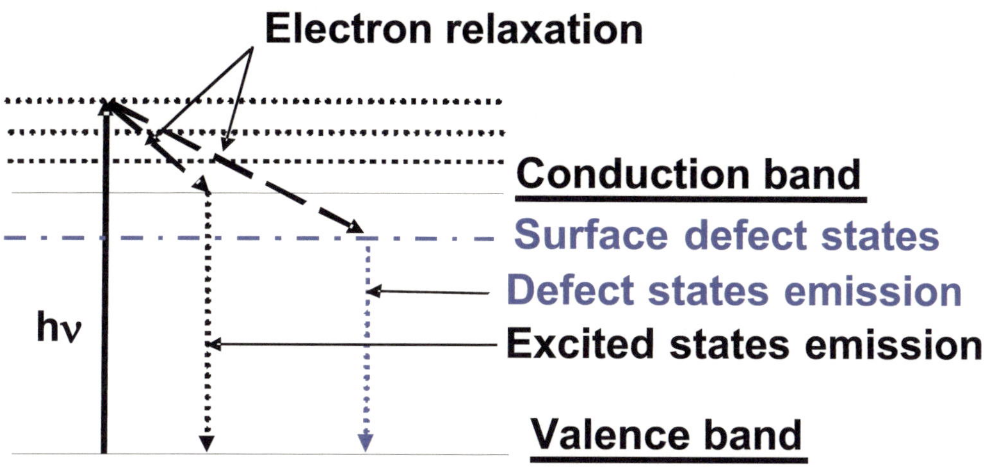

Fig. 1. Schematic diagram illustrating basic emission properties of CdSe quantum dots.

3. Synthesis of CdSe nanoparticles in aqueous phase

3.1 Syntheses without ligand or with weakly-coordinating ligands

Many properties of CdSe, such as electrical and optical properties, are mainly controlled by the particles size, size distribution, phase, and morphology. Recent efforts to synthesize nanostructures with well-defined geometrical shapes and organize them as two- or three-dimensional assemblies have further expanded the possibility of developing materials for optoelectronic devices and sensing. The first part of this chapter will be quite entirely devoted to the preparation of low-dimensional CdSe nanomaterials as extensions of the zero-dimensional CdSe quantum dots that will be described in details in the following paragraphs.

Solvothermal synthetic methods have initially been developed by using organic solvents such as pyridine or ethylene diamine at high temperature for the preparation of CdSe nanoparticles (Yu et al., 1998). Treatment of cadmium oxalate CdC_2O_4 with Se at 140°C in chelating solvents such as ethylene diamine, diethylene triamine or triethylene tetramine afforded CdSe nanorods in pure hexagonal phase with lengths up to 100-500 nm and diameters ranging from 6 to 20 nm (Figure 1).

$$CdC_2O_4 + Se \xrightarrow{140°C} CdSe + 2\,CO_2$$

Fig. 2. Synthesis of CdSe nanoparticles from cadmium oxalate and Se(0)

The morphology of CdSe crystallites can be controlled by the solvent. Uniform disklike particles with a size of 40 nm were prepared using the same synthetic protocol but performing the reaction in pyridine as solvent.

By changing the synthetic route, the same authors demonstrated one year later that spherical CdSe nanoparticles with an average diameter of 7 nm could be prepared in the ethylene diamine using $CdCl_2$ and Se associated to sodium metal (Figure 3). Crystalline

CdSe nanoparticles produced by this solvothermal method have a hexagonal structure. Upon excitation at 488 nm, a sharp emission was observed at 600 nm indicating a quantum confinement (Xie et al., 1999).

$$CdCl_2 + Se + 2\,Na \xrightarrow[6\,h]{120°C} CdSe + 2\,NaCl$$

Fig. 3. Synthesis of CdSe nanoparticles from CdCl2, Se(0) and Na(0)

The synthesis in aqueous media has further been investigated because it has been identified as a cheaper and cleaner route to CdSe nanoparticles.

CdSe nanoparticles with sizes ranging from 70 to 100 nm can simply be produced by reacting metallic Cd with Se in an autoclave at 180°C for 24 hours. Cd(0) was found to react with H_2O at high pressure and temperature to generate $Cd(OH)_2$ and H_2. Dissolved Se is reduced by H_2 to produce Se^{2-}, which reacts with $Cd(OH)_2$ to give CdSe nanocrystals (Figure 4). Interestingly, CdSe nanocrystals produced at that high temperature have a cubic phase structure (Peng et al., 2001).

$$3\,Cd + 6\,H_2O \longrightarrow 3\,Cd(OH)_2 + 3\,H_2$$
$$3\,Cd(OH)_2 + 3\,Se + 3\,H_2 \longrightarrow 3\,CdSe + 6\,H_2O$$

Fig. 4. Synthesis of CdSe nanoparticles from Cd(0) and Se(0)

Ethylene diamine can also be used as surface ligand for the stabilization of CdSe QDs prepared under hydrothermal conditions (100°C, 10 hours, autoclave) from $CdCl_2$ and sodium selenosulfate Na_2SeSO_3. Cubic zinc blende CdSe nanoparticles with an average diameter of 12 nm were further used for the preparation of CdSe-oligonucleotide probe. Because the surface ethylene diamine ligand is positively charged at neutral pH, these dots were found to associate with the negatively charged phosphate skeleton of the DNA molecule (Huang et al., 2009).

Triethanolamine (TEA) was also recently used as capping ligand to prepare CdSe nanoparticles. Selenide Se^{2-}, produced by reaction of Se with sodium borohydride $NaBH_4$, reacts at room temperature with $CdCl_2$ in the presence of TEA (Cd/TEA = 1/20) to yield CdSe nanocrystals with an average diameter of ca. 8 nm. The amine group of TEA binds to surface Cd atoms, while the OH groups provide hydrophilicity (Dlamini et al., 2011).

High quality CdSe nanorods and fractals can be prepared from Na_2SeO_3 and $Cd(NO_3)_2$ at temperatures varying from 100 to 180°C and using hydrazine N_2H_4 as reductant (Figure 5). Na_2SeO_3 is first reduced into Se(0) by N_2H_4. Through a disproportiotionating reaction in the basic solution, Se is next converted into Se^{2-} and SeO_3^{2-}. Se^{2-} finally reacts with $Cd(NH_3)_4^{2+}$ to yield CdSe (Peng et al., 2002).

$$2\,Cd(NO_3)_2 + 2\,Na_2SeO_3 + 3\,N_2H_4 \xrightarrow{NH_3.H_2O} 2\,CdSe + 3\,N_2 + 4\,NaNO_3 + 6\,H_2O$$

Fig. 5. Synthesis of CdSe nanoparticles from $Cd(NO_3)_2$ and Na_2SeO_3 using hydrazine as reducing agent

CdSe fractals with wurtzite structure were produced using $NH_3.H_2O$ as complexing agent and operating at low temperature. Using the stronger complexing agent EDTA and working at higher temperature afforded nanorods with pure zinc blende phase.

Close results were obtained using cetyltrimethylammonium bromide (CTAB) as surfactant and operating at 180°C for 10 hours. Dendritic structures and nanoparticles are produced in the absence or at low CTAB concentration, while high CTAB concentration favored the production of nanorods with wurtzite structure (diameter of 40-60 nm and lengths between 200 and 500 nm). In that surfactant-assited hydrothermal system, CTAB adsorbs to the surface of newly nucleated nanoparticles. As a result, the growths on same faces were shut down, and the c-axis direction was kept as preferential orientation (Chen et al., 2005).

CdSe nanochains can be produced through a hydrothermal method using $Cd(OH)_2$ nanoflakes as sacrificial template. $Cd(OH)_2$ was first produced by basic hydrolysis of $Cd(NO_3)_2$. the quite perfect hexagonal nanoflakes obtained have a side length of ca. 200 nm. These nanoflakes were further selenized using Na_2SeSO_3 at 160°C for 10 h (Figure 6).

$$SeSO_3^{2-} + 2\,OH^- \longrightarrow Se^{2-} + SO_4^{2-} + H_2O$$

$$Cd(OH)_2 + Se^{2-} \longrightarrow CdSe + 2\,OH^-$$

Fig. 6. Preparation of CdSe nanochains from $Cd(OH)_2$ nanoflakes

Wurtzite CdSe nanoparticles produced arranged in open nanochains, which maintain the external skeleton of $Cd(OH)_2$ nanoflakes. The preferential reaction of Se^{2-} with the edges of $Cd(OH)_2$ nanoflakes where there is a large curvature and a high chemical reactivity, was used to explain the formation of nanochains (Chen et al., 2006).

$Cd(OH)_2$ nanowires prepared by treatment of $CdSO_4$ with NaOH followed by hydrothermal treatment (180°C, 12 h) in the presence of sodium dodecylbenzene sulfonate (SDBS) were also used for the preparation of CdSe nanoparticles. Selenization of these nanowires was accomplished by treatment with Se powder and hydrazine in ethylene diamine at 180°C for 20 hours. The 30 nm diameter CdSe particles in cubic phase were evaluated as photocatalyst for the degradation of organic dyes such as Safranine T and Pyranine B. Under irradiation of 365 nm UV light, CdSe nanoparticles exhibit higher photocatalytic activity than CdS, CdO and $Cd(OH)_2$ (Li et al., 2009)

Using *N,N*-dimethyloctadecylammonium bromide cadmium acetate (C_{18}-Zn) and Na_2SeSO_3, the ionic liquid lithium *bis*((trifluoromethyl)sulfonyl)amide ($LiN(Tf)_2$) as template and performing the reaction for two days at 180°C, CdSe nanorods with a diameter of 10 nm and a length of 30-50 nm could be prepared. These nanorods tended to auto-organize into chain-like patterns up to 1 µm or more in length and exist in a mixture of cubic and hexagonal crystalline phase forms. Interestingly, in the absence of ionic liquid, 10-30 nm wide CdSe nanoparticles in pure cubic phase were obtained (Tao et al., 2010). The surface photovoltaic response of CdSe nanorod-chain assemblies was observed at wavelengths in the range 320-800 nm and was found to be 40 times higher compared with that of CdSe nanocrystals. Moreover, the intensity of the photocurrent of CdSe nanorods was about 10 times that of CdSe nanocrystals and these nanorods have therefore great promise as photoelectric sensor.

Four different morphologies (taper microrods, nanotetrapods, fringy nanostructures and fasciculate nanostructures) were recently prepared from $Cd(NO_3)_2$, Na_2SeO_3 and ethylene diamine tetraacetic acid tetrasodium salt (EDTA) as both a chelating agent and a reductant. At elevated temperature, SeO_3^{2-} is first reduced by EDTA to Se(0), which disproportionates at basic pH to give Se^{2-} and SeO_3^{2-}. The authors demonstrate that the important factors that influence the morphology of CdSe crystals are the concentration of NaOH, the reaction temperature and, to lesser extent, the concentration of SeO_3^{2-}. At low NaOH concentration, the formation of Se^{2-} is slow, which favors the anisotropic nucleation and growth of the CdSe structure along the c-axis (formation of rods). With an increased quantity of NaOH, the concentration of Se^{2-} becomes higher, providing more CdSe building units that grow on nanorods (Liu et al., 2010).

A solvothermal route using a mixture of triethylene tetramine (TETA) and de-ionized water was also developed for the synthesis of CdSe microspheres with an average diameter of 5 µm. $Cd(NO_3)_2$ and Se were used as precursors and TETA played the role of reducing agent and surfactant. CdSe particles obtained after the solvothermal reaction performed at 180°C for 12 hours were further treated at 580°C for 4 hours to get pure CdSe in wurtzite structure. When the reaction was conducted with hydrazine in place of TETA as reductant, CdSe nanorods and dendrites were obtained (Yang et al., 2010).

Two papers have also been dedicated to ultrasonic-assisted synthesis of CdSe nanoparticles. The implosive collapse of bubbles generated from acoustic cavitation generates hot spots with transient temperatures of ca. 5000 K, pressures over 1800 kPa, and cooling rates in excess 10^{10} $K.s^{-1}$. Hence, sonochemistry is becoming a very attractive method for the preparation of nanomaterials.

Depending on the sonication conditions, amorphous or wurtzite CdSe nanocrystals can be produced using $Cd(NO_3)_2$, Na_2SeO_3, NH_3 and N_2H_4 as starting materials (Figure 7).

$$3\,SeO_3^{2-} + 3\,N_2H_4 \longrightarrow 3\,Se + 3\,N_2 + 3\,H_2O + 6\,OH^-$$

$$3\,Se + 6\,OH^- \longrightarrow 2\,Se^{2-} + SeO_3^{2-} + 3\,H_2O$$

$$2\,Se^{2-} + 2\,[Cd(NH_3)_4]^{2+} \longrightarrow 2\,CdSe + 8\,NH_3$$

$$2\,SeO_3^{2-} + 3\,N_2H_4 + 2\,[Cd(NH_3)_4]^{2+} \longrightarrow 2\,CdSe + 8\,NH_3 + 3\,N_2 + 6\,H_2O$$

Fig. 7. Ultrasonic-assisted synthesis of CdSe nanocrystals

Particles thus prepared have small diameters (ca. 5-6 nm) but tend to aggregate. Amorphous CdSe nanoparticles have a luminescence 5-times higher compared to hexagonal CdSe (excitonic emission at 556 nm and broad emission centered near 670 nm). This luminescence was found to be pressure sensitive due to a solid-solid phase transition (Ge et al., 2002).

A sonochemical route was also developed for the preparation of CdSe hollow spherical structures (Figure 8).

$$H_2O \longrightarrow H^{\cdot} + OH^{\cdot}$$

$$2\,H^{\cdot} + SeSO_3^{2-} \longrightarrow Se^{2-} + 2\,H^+ + SO_3^{2-}$$

$$Cd(OH)_2 + Se^{2-} \longrightarrow CdSe + 2\,OH^-$$

Fig. 8. Generation of Se^{2-} under sonication and preparation of CdSe nanocrystals

$Cd(OH)_2$ produced by reaction of $CdCl_2$ with aqueous NH_3 or triethylamine generates CdSe nanoparticles upon reaction of H$^{\cdot}$ with sodium selenosulfate. Amorphous $Cd(OH)_2$, which acts as an *in situ* template, was found to direct the growth of the primary CdSe nanoparticles formed and their assembly into hollow spheres of ca. 120 nm in diameter (Zhu et al., 2003).

Finally, γ-irradiation has been used to prepare CdSe nanocrystals. Electrons or H$^{\bullet}$ radicals produced after γ-ray irradiation of water at room temperature reduce SeO_3^{2-} into Se^{2-}, which could react with $[Cd(NH_3)_4]^{2+}$ to produce cubic CdSe nanocrystals with an average diameter of 4 nm (Figure 9) (Yang et al., 2003).

$$SeO_3^{2-} + 4\,e^- \text{ (or 4 H}^{\cdot}\text{)} \longrightarrow Se + 2\,e^- \text{ (or 2 H}^{\cdot}\text{)} \longrightarrow Se^{2-}$$

$$Se^{2-} + [Cd(NH_3)_4]^{2+} \longrightarrow CdSe + 4\,NH_3$$

Fig. 9. γ-Irradiation assisted synthesis of CdSe nanoparticles

3.2 Citrate-capped CdSe QDs

The weakly coordinating trisodium citrate ligand (Figure 10), which stabilizes nanoparticles through electrostatic interactions, has numerous times been used to cap CdSe QDs prepared in aqueous solution.

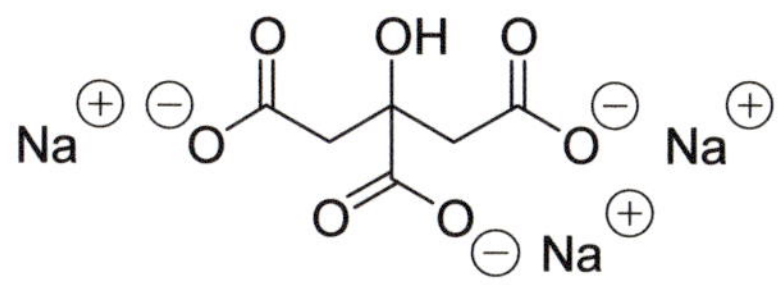

Fig. 10. Chemical structure of trisodium citrate

First syntheses of citrate-capped CdSe QDs were performed under microwave irradiation because the microwave process provides a uniform thermal activation that is ideal for nuclei formation and growth of nanoparticles.

Citrate-capped CdSe QDs were first prepared by Rogach *et al.* using $Cd(ClO_4)_2$ and *N,N*-dimethylselenourea at pH = 9 (Figure 11).

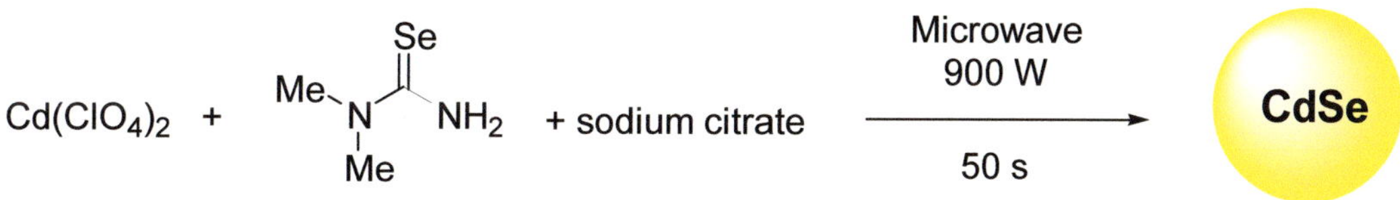

Fig. 11. Synthesis of CdSe nanoparticles from $Cd(ClO_4)_2$ and *N,N*-dimethylselenourea under microwave irradiation

Using a Cd/Se ratio of 4/1 affords CdSe nanocrystals with ca. 4.0 nm diameter. By varying the Cd-to-Se ratio (from 16:1 to 4:1), it is also possible to prepare CdSe QDs of different sizes (1-5 nm). These nanocrystals exhibit excitonic emission located between 550 and 600 nm but also trapped emissions between 650 and 800 nm. The PL QYs are weak (0.1 0.15%) but can be enhanced to 4.2% by introduction of a CdS shell at the periphery of

QDs by decomposition of thioacetamide. Correspondingly, the trapped emission of bare CdSe QDs is strongly depressed for the core/shell CdSe/CdS nanocrystals because the shell passivates nanoparticles surface defects, thus leading to the localization of photoexcited charge carriers in the CdSe core. The surface of these nanocrystals was successfully modified with 2-mercaptopropyltrimethoxysilane. By addition of sodium silicate, CdSe or CdSe/CdS QDs could be incorporated into silica spheres of 40-80 nm diameter (Figure 12) (Rogach et al., 2000).

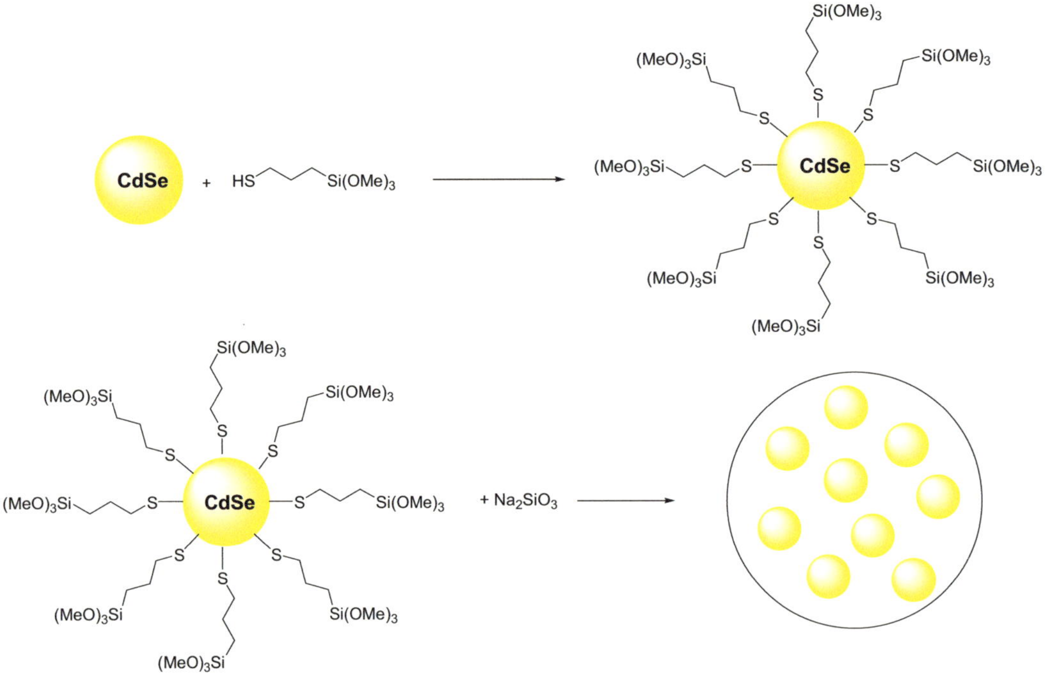

Fig. 12. Preparation of silica nanospheres containing CdSe QDs

The same group reported one year later a strong increase of PL QY of CdSe QDs through photoetching. Exposure of CdSe/CdS QDs dispersions in water to ambient light for several days (intensity 0.12 mW) afforded dots with PL QYs reaching values of 25-45%, while no increase in PL was observed for the samples stored in the dark. The photoetching was accompanied by a blue shift of the PL peak, and the blue-shift was more marked for the largest particles (up to 30 nm for yellow-emitting CdSe QDs) (Wang et al., 2003).

The photooxidation of the QDs surface in the presence of oxygen was demonstrated to play a crucial role in the photoactivation process. Electrons propulsed to the conduction band upon light-activation react with O_2 to generate reactive-oxygen species like $O_2^{\bullet -}$, while holes trapped on the surface oxidize Se into SeO_2. This results in the gradual erosion of the unwanted topographic features on the surface and in smooth, which yields highly luminescent nanoparticles where the nonradiative decay of excitons no longer dominates. It is finally worth mentioning that the photoactivation was found to be the more efficient for CdSe/CdS capped by a thin silica shell. Indeed, the porosity of the silica shell results in the enhanced adsorption of O_2 molecules from the solution which are therefore readily available for the photocorrosion reactions and therefore for the photoactivation process (Wang et al., 2004).

Using the same starting materials ($Cd(ClO_4)_2$, $Me_2NCSeNH_2$), microwave irradiation, and a [Cd]/[Se] ratio of 4 permits to finely tune the size of the dots with the temperature, while a [Cd]/[Se] ratio of 8 allows to prepare QDs with the highest PL QY. By varying the temperature from 60 to 180°C, nanocrystals with diameters ranging from 2.5 to 4.0 nm were readily prepared. Of course, CdSe QDs prepared at high temperature possess the best optical properties (no trapped-state emission) due to their good crystallinity and narrowing of size distribution. The PL QY measured for bare CdSe QDs prepared at 180°C with [Cd]/[Se] of 8 was found to be 9.9%. PL lifetimes were also increased for CdSe QDs prepared at high temperature. To passivate the unsaturated surface dangling bonds, sequester excitons within the core and thus increase PL QYs to 20-40%, a ZnS shell was successfully introduced at the periphery of CdSe QDs by thermal decomposition of $Zn(ClO_4)_2$ and thioacetamide CH_3CSNH_2 (Figure 13) (Han et al., 2010).

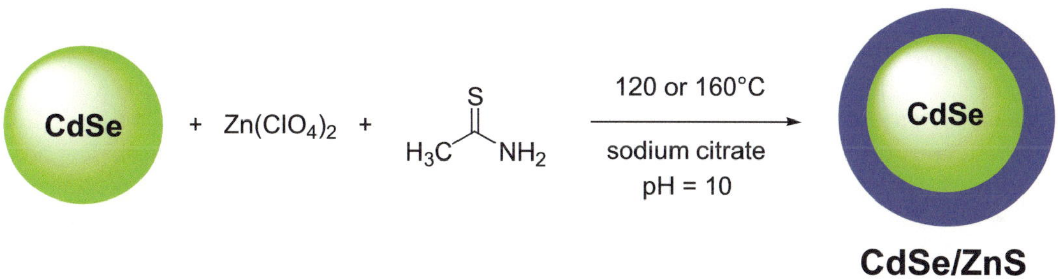

Fig. 13. Preparation of core/shell CdSe/ZnS QDs under microwave irradiation

The first high temperature synthesis of citrate-capped CdSe QDs was described in 2007 by Williams *et al.* Mixing $Cd(ClO_4)_2$, *N,N*-dimethylselenourea and sodium citrate at room temperature followed by heating to 200°C for 15 min. produced CdSe nanocrystals. Using a [Cd]/[Se] ratio of 8/1 and a [Cd]/[Citrate] ratio of 1.04, the PL QY of the CdSe QDs was found to be modest (1.5%) but could be increased to 7% by adding a CdS shell through thermal decomposition at 200°C of thioacetamide. The authors demonstrated that the mean diameter of the nanocrystals can be tuned by changing the process variables. The particle size increased with increasing reaction time, temperature, and the Cd/Se ratio. Surprisingly, increasing the loading in citrate had the same effect and this may indicate that nucleation and growth processes in high-temperature liquid water differ from classical CdSe QDs syntheses. The mean diameter of CdSe dots was finally found to decrease with increasing pH (from 9 to 11) because at high pH the coordination between the citrate ligand and Cd surface atoms strengthens and Ostwald ripening is therefore slowed (Williams et al., 2007).

The same authors reported two years later an improved synthetic procedure for the preparation of CdSe QDs based on the rapid injection at 200°C of selenourea to cadmium perchlorate. Using that methodology, smaller nanoparticles were produced (3.5 versus 5.0 nm using a Cd/Se/Citrate ratio of 8/1/9.5). Moreover, the widths of the PL emission peaks were found to be smaller than those obtained when mixing the reagents at room temperature. Without any post-synthetic treatment, citrate-capped CdSe nanoparticles have a PL QY over 5%. The dependence of the PL emission peak position on the Cd:Se ratio during the synthesis was also demonstrated. Contrary to results obtained using the cold-precursor loading method, a decrease in nanoparticle diameters as the Cd:Se molar ratio increased was observed (Williams et al., 2009).

Photo-irradiation with a high pressure mercury lamp at room temperature was also developed to prepare 4 nm-sized CdSe nanocrystals from $Cd(OAc)_2$, Se and Na_2SO_3 (Figure 14) using sodium citrate and cetyltrimethylammonium bromide used as capping agents. The optical properties of the CdSe nanocrystals have not been reported probably due to their important size polydispersity (Yan et al., 2003).

$$2\,SO_3^{2-} + h\nu \longrightarrow \left[2\,SO_3^{-}\right] + 2\,e^-$$

$$Se + 2\,e^- \longrightarrow Se^{2-}$$

$$Cd^{2+} + Se^{2-} \longrightarrow CdSe$$

Fig. 14. Photo-assisted synthesis of CdSe nanoparticles

Because classical syntheses of CdSe QDs in aqueous solution at temperatures below 100°C generally afford nanocrystals with poor PL QYs (< 0.15%), laser-assisted synthesis has been developed. Heating a mixture of $Cd(NO_3)_2$ and *N,N*-dimethylselenourea in the presence of sodium citrate while irradiating with a Nd:YAG laser at 532 nm affords CdSe QDs emitting between 561 and 590 nm depending on the temperature used during the synthesis. UV-vis spectroscopy monitoring of the reaction demonstrated that photoetching or photooxidation of the surface of the CdSe QDs dominates the growth of the nanocrystals during the first half-hour of the reaction before nanoparticles growth through Ostwald ripening takes control. To improve PL QYs, a CdS shell was first introduced at the periphery of CdSe cores using thioacetamide as sulphur source and the crude solution illuminated for 24 h with a 100 W Xe-Hg lamp to photopassivate the core/shell CdSe/CdS QDs. The PL QY of the obtained QDs reaches 60% and the full-width-at-half-maximum (fwhm) of the PL emission peak is below 35 nm, indicating a narrow size distribution.

By varying the [Cd]/[Se] ratio from 2 to 12, this simple method allows the preparation of CdSe/CdS QDs with sizes between 3.7 and 6.3 nm that emit in a wide range in the visible region (green to red) (Lin et al., 2005).

Illumination with ambient light has also been used to improve the PL QY of core/shell CdSe/CdS QDs. CdSe cores were first prepared by bubbling H_2Se, generated from NaHSe and H_2SO_4, into a solution of $CdCl_2$ and sodium citrate adjusted to pH = 9. CdSe/CdS QDs were prepared from the crude citrate-capped CdSe QDs by bubbling H_2S (Figure 15).

By varying the temperature of the reaction medium or the heating time during the preparation of the cores from 20 to 94°C, CdSe QDs with cubic structure and emitting between 500 and 570 nm could be prepared. After introduction of the CdS shell and upon illumination for 30 days, core/shell CdSe/CdS QDs with diameters below 4.0 nm, with PL QYs over 20% and with narrow luminescence band (fwhm of ca. 40 nm) were prepared (Deng et al., 2006).

Because of the higher affinity of Cd for thiol compounds than for carboxylate functions, citrate-capped CdSe QDs can easily be modified with thiol-bearing compounds. This strategy was successfully used for the surface modification of core CdSe and core/shell CdSe/CdS QDs with monothiolated β-cyclodextrin (mSH-β-CD). Core CdSe were classically prepared by reaction of $Cd(ClO_4)_2$ with *N,N*-dimethylselenourea at 100°C and the CdS shell introduced at the periphery of citrate-capped CdSe QDs by thermal decomposition of thioacetamide. Both dots were modified by the cyclodextrin simply by adding mSH-β-CD to the citrate-capped CdSe core QDs at 75°C and stirring for 6 to 24 hours under N_2 (Figure 16).

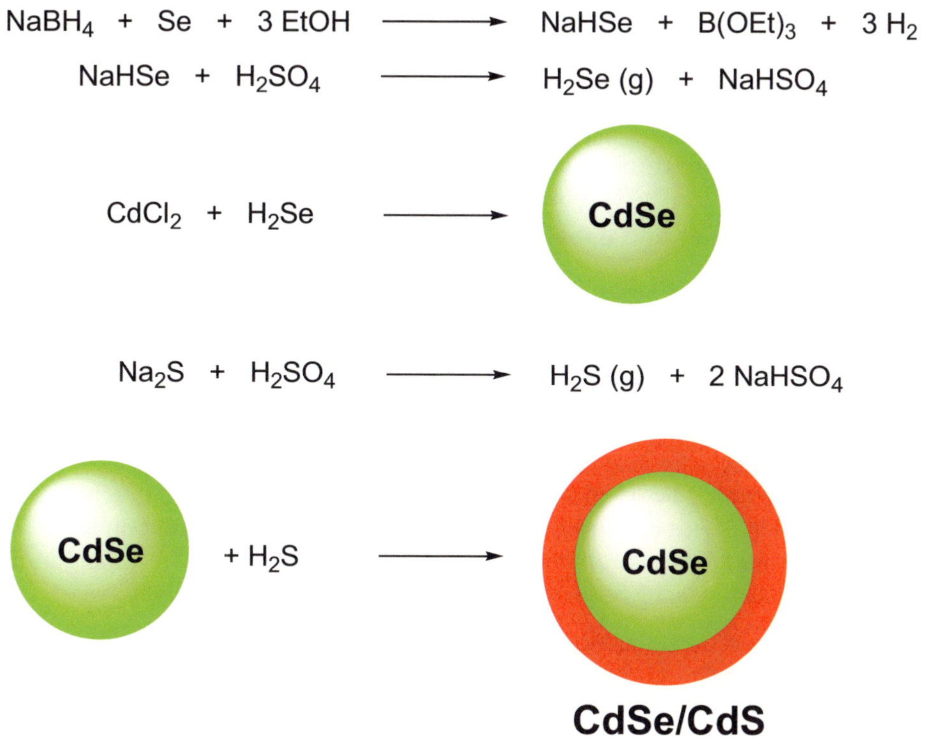

Fig. 15. Synthesis of core/shell CdSe/CdS QDs

Only band-edge emission at 597 nm was observed for CdSe/CdS QDs, while band-edge and trap-state emissions, at 550 and 700 nm respectively, could be detected for CdSe QDs. The PL QYs of these dots were found to be modest (0.65% and 3.7% for CdSe and CdSe/CdS, respectively) but could be markedly improved simply by exposing water dispersions of the nanocrystals to room light over 1 week in sealed tubes. PL QYs increased from the original 0.65% and 3.7% to 7.8% and 46.2%, respectively for CdSe and CdSe/CdS. The numerous hydroxyl groups of surface-bound mSH-β-CD conferred to the QDs high water dispersibility at pH values between 5 to 9 and high ionic strength media. The surface-anchored mSH-β-CD was also found to retain its host capability for small hydrophobic molecules in aqueous media. The fluorescence of the β-CD-modified CdSe core QDs could be selectively tuned by introducing different redox-active substrates (e.g. fluorescence quenching with ferrocene and fluorescence enhancement with benzoquinone) and these nanocrystals have therefore great promise for sensing (Palaniappan et al., 2006).

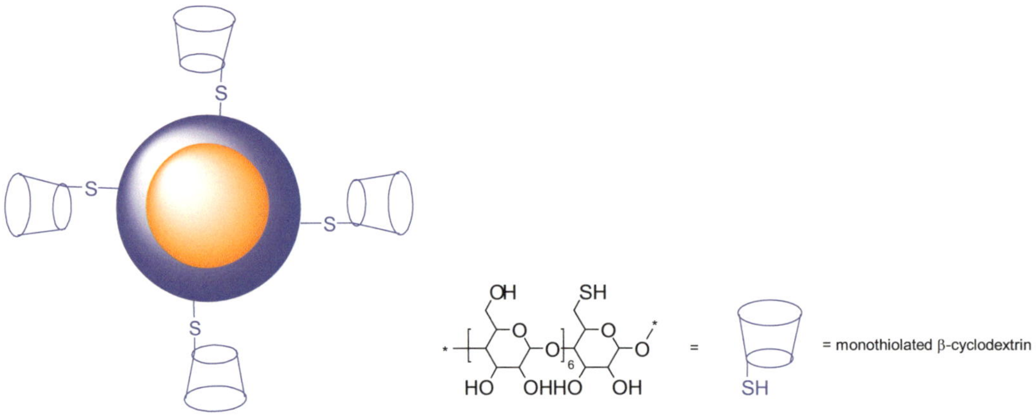

Fig. 16. structure of cyclodextrin-capped CdSe/Cds QDs

3.3 Thioacids, thioamines and thioalcohols as stabilizers

Organic molecules with both thiol and carboxylic acid, amine or alcohol functional groups have been widely adopted as capping molecules for CdSe nanoparticles in aqueous phase. The thiol group coordinates to surface Cd atoms, whereas carboxylate, ammonium or alcohol groups contribute to the electrostatic stabilization of the colloidal nanoparticles as well as to their further surface modification for various applications.

CdSe QDs stabilized by 4-mercaptobenzoic acid and possessing an average diameter of ca. 2 nm can easily be prepared by reaction of $Cd(OAc)_2$ with sodium hydrogenoselenide in DMF at room temperature. These nanoparticles exhibit a strong exciton absorption at 430 nm and were successfully used for the preparation of poly(vinylpyridine) (PVP)/CdSe thin films by layer-by layer deposition using hydrogen bond interaction (Hao et al., 2000).

Parallel methods based on an aqueous route were also successfully developed using various thioacids or thioalcools as capping ligands. A paper published in 2000 reviews comprehensively the early work devoted to the aqueous-phase synthesis of thiol-capped II-VI semiconductor nanocrystals (Eychmüller et al., 2000).

Rogach *et al.* described in 1999 the first aqueous preparation of CdSe QDs using thioacids (thioglycolic acid (TGA) or thiolactic acid) or thioalcohols (2-mercaptoethanol, 1-thioglycerol) as stabilizers (Rogach et al., 1999). Reactions were conducted at 100°C using a Cd^{2+}/Se^{2-}/stabilizer ration of 2/1/4.8 based on the conditions used for the preparation of CdS and CdTe nanoclusters (Rogach et al., 1996; Vossmeyer et al., 1995). The growth of nanocrystals via Ostwald ripening using thioalcohols as capping agents was found to be significantly reduced (1.4-2.2 nm) compared to those prepared with TGA (2.1-3.2 nm). Photoluminescence of all the QDs prepared was found to originate from deep surface traps (from 600 nm for thioalcool stabilized CdSe QDs to near IR for TGA-capped CdSe QDs) and PL QYs were weak (< 0.1%).

A deep trap emission centered at 560 nm originating from incompletely surface-passivated QDs was also observed using Na_2Se instead of NaHSe and performing the reaction with 3-mercaptopropionic acid (3-MPA) as capping ligand. The authors studied the fluorescence quenching of these 1.5-2.0 nm-sized CdSe QDs by various gold nanoparticles and demonstrated that the quenching rate of Au nanorods is more than 1000 times higher stronger than that of Au nanospheres (Nikoobakht et al., 2002).

By varying the Cd^{2+}/NaHSe/3-MPA ratio (optimal ratio determined = 8/1/20) and performing the reaction at pH = 9.5 under microwave irradiation (temperature = 140°C), the low energy band of emission from trap states could markedly be reduced. Alloyed CdSe(S) QDs with a gradient of sulfur concentration from the core to the surface were formed due to the decomposition of the 3-MPA ligand under microwave irradiation and release of sulfide ions S^{2-} in the reaction medium. The PL QY of the CdSe(S) QDs was found to be 25% in water and full-width-at-half maximum was about 28 nm (Qian et al., 2005).

Small-sized CdSe QDs (diameter of ca. 2 nm) were also produced under high-intensity ultrasonic irradiation. $H^{\bullet}$ radicals formed during the sonochemical process reduce Na_2SeSO_3 into Se^{2-} which reacts with $CdCl_2$ and thioglycolic acid at pH = 11 to produce CdSe@TGA QDs (Figure 17). Using a Cd/Se ratio of 0.67, the first excitonic peak was located at ca. 450 nm. The photoluminescence properties of the dots were not described by the authors (Han et al., 2006).

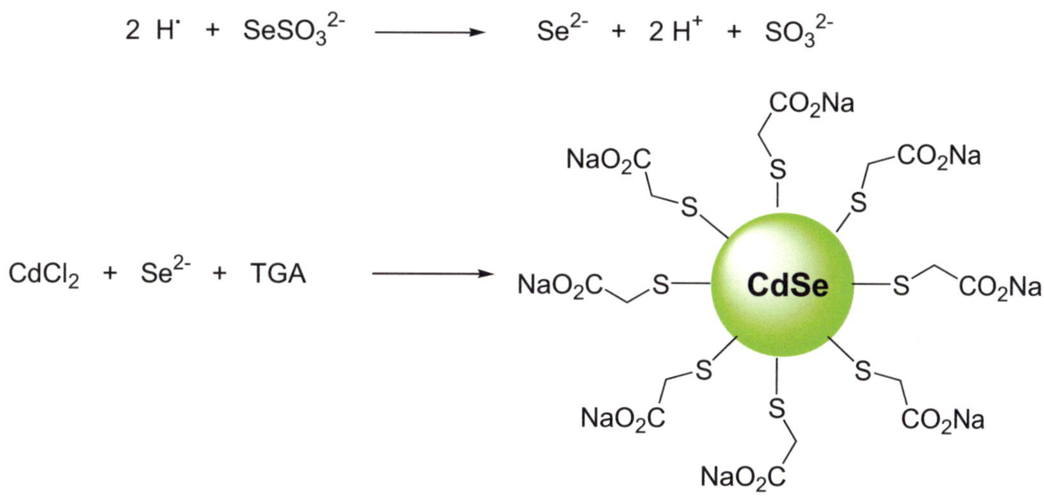

Fig. 17. Synthesis of CdSe@TGA QDs under ultrasonic irradiation

The amphiphilic 11-mercaptoundecanoic acid (MUA) was also used to disperse CdSe QDs in aqueous solution. The synthesis performed under standard conditions (Cd/Se/MUA = 2/1/4.8; pH = 11) in refluxing water for 2 h afforded CdSe with an average diameter of ca. 6 nm. A sharp excitonic peak was observed at 327 nm but the emission, centered at 550 nm, was found to be broad. CdSe@MUA QDs were successfully used as specific fluorescent probe for lysozyme (limit of detection of 0.115 $mg.mL^{-1}$) (Zhong et al., 2006).

The same synthetic protocol was used for the preparation of CdSe@TGA QDs emitting at 525 nm. The dots could successfully be linked with bacterial cells (*Escherichia coli* and *Staphylococcus aureus*) using 1-ethyl-3-(3-dimethylaminopropyl) carbodiimide) (EDC) as coupling reagent (Figure 18). A sensitive and rapid fluorescence method was developed for counting these bacteria (detection of 10^2-10^7 CFU/mL total count of *E. coli* and *S. aureus* in 1-2 h; limit of detection: 10^2 CFU/mL) (Xue et al., 2009).

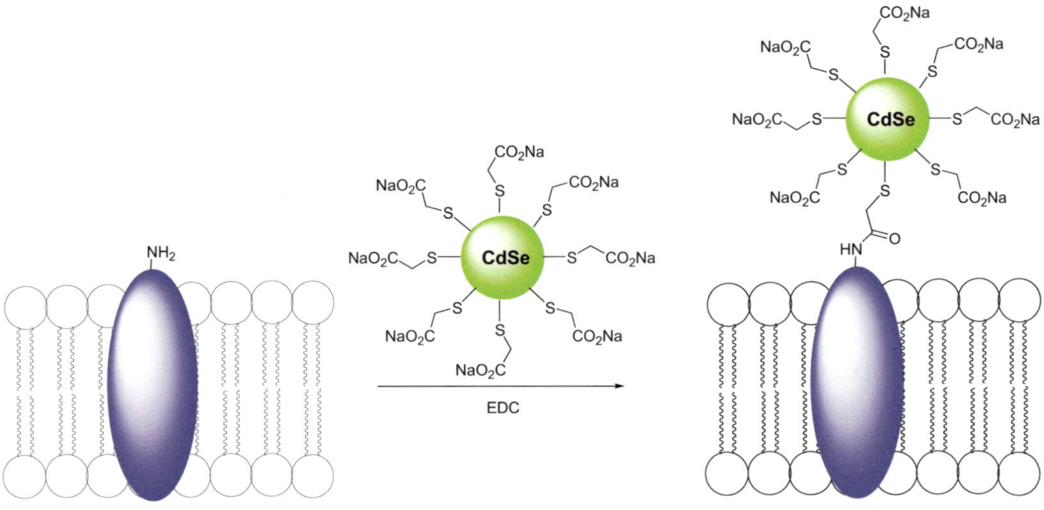

Fig. 18. Principle of EDC-mediated bacterial membrane coupling with CdSe@TGA QDs

Our group has also studied the influence of the thioalkyl acid during the aqueous synthesis of CdSe QDs and demonstrated that the particle growth and thus the photophysical properties of the nanocrystals are related to the structure of the capping ligand. The growth at 100°C of CdSe QDs was markedly decreased with 6-mercaptohexanoic acid (MHA) and MUA compared to MPA yielding particles with diameters of 2.8, 2.5 and 1.9 nm, respectively for MPA, MHA, and MUA ligands. The PL emission of these QDs was dominated by low-energy bands arising from the recombination of trapped charge carriers (Figure 19). By changing the Cd/Se/MPA from 2/1/4.8 to 8/1/8 and performing the reaction under pressure at 150°C for 1 hour in a Teflon-lined stainless steel autoclave, good quality (PL QY of 12%) green-emitting alloyed CdSe(S)@MPA QDs could be prepared (Aldeek et al., 2008).

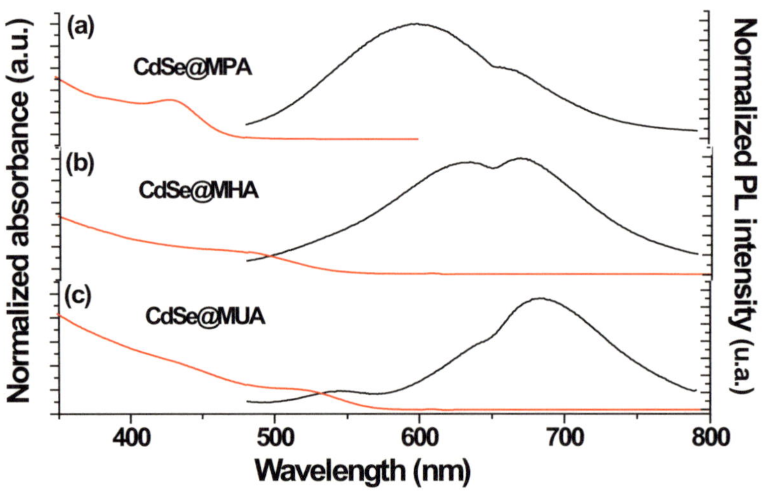

Fig. 19. UV-vis absorption (red) and PL (black) spectra of (a) MPA-, (b) MHA- and, (c) MUA-stabilized CdSe QDs prepared at 100°C for 15 h. PL spectra were recorded with an excitation at 400 nm.

The PL QY and the photostability of the green-emitting CdSe(S)@MPA QDs could be improved by epitaxial overcoat of a ZnO shell on the outerlayer of alloyed CdSe(S)@MPA QDs. Using Zn/Cd ratios of 0.2 or 0.4, the PL QYs of core/shell CdSe(S)/ZnO@MPA QDs reached 20 and 24%, respectively (Figure 20) (Aldeek et al., 2011).

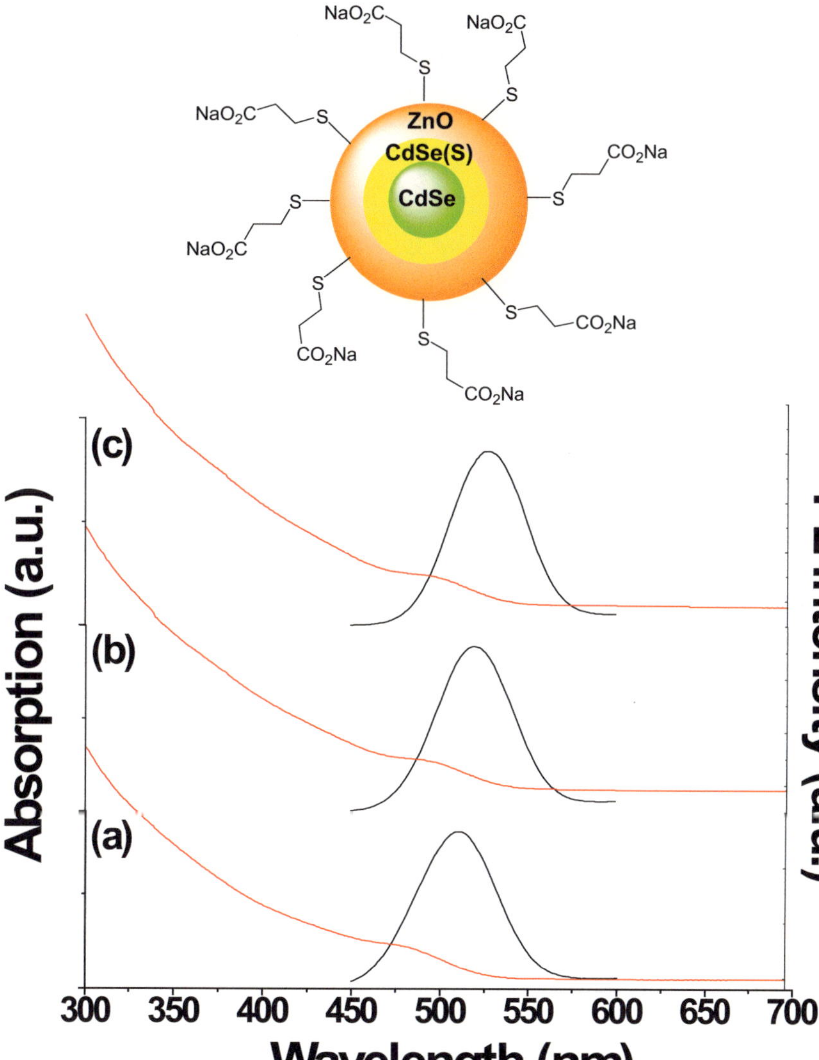

Fig. 20. UV-vis absorption (red) and PL (black) spectra of (a) core CdSe(S) QDs, (b) CdSe(S)/ZnO QDs (Zn/Cd = 0.2), and (c) CdSe(S)/ZnO QDs (Zn/Cd = 0.4).

The potential of these highly luminescent CdSe(S)/ZnO QDs for bioimaging applications was demonstrated by the labelling of *Schewanella oneidensis* biofilms (Figure 21).

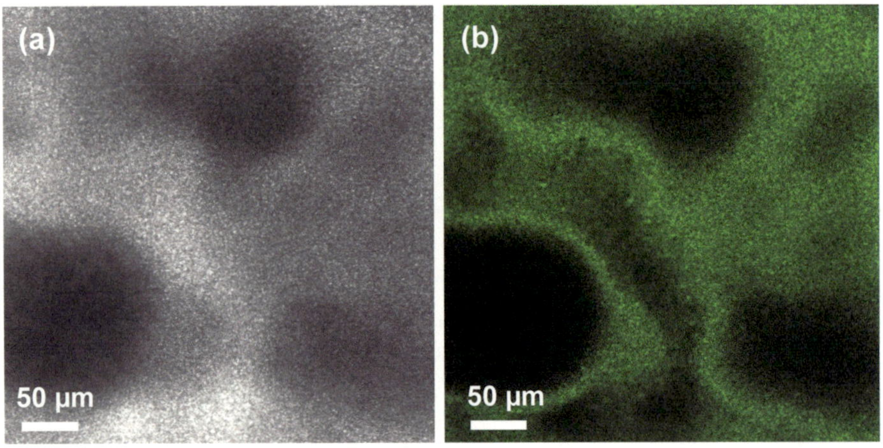

Fig. 21. Confocal microscopy images of a *S. oneidensis* biofilm treated with CdSe(S)/ZnO QDs: (a) transmission image, and (b) corresponding *XY* plane fluorescence image. Confocal microscopy images were obtained with laser excitation at 405 nm.

Hydrazine hydrate/Se complex as the source of selenium was also developed to prepared high quality CdSe@3-MPA QDs under ambient conditions (Figure 22).

$$Se + NH_2NH_2 \longrightarrow Se^{2-}(NH_2NH_2)^{2+}$$

$$CdCl_2 + Se^{2-}(NH_2NH_2)^{2+} \longrightarrow CdSe + N_2 + 2\,HCl + H_2$$

Fig. 22. Synthesis of CdSe QDs using a hydrazine/Se complex and $CdCl_2$

Using a Cd/Se ratio of 8/1, the size of the nanocrystals could readily be tuned from 1.6 to 3.5 nm by controlling the annealing time of the colloid at 100°C and nanocrystals emitting from blue to red were prepared (PL QY up to 40%) (Kalasad et al., 2009).

Mercaptosuccinic acid (MSA), bearing two carboxylic acid functions, was also used to stabilize CdSe QDs. The 4.2 nm-sized CdSe@MSA nanocrystals obtained after 2 hours heating at 90°C have a cubic structure structure and emit at ca. 540 nm. The fluorescence of these dots was found to be sensitive to Cu^{2+} ions (limit of detection 3.4 $nmol.L^{-1}$) but not to other physiologically important cations (Zn^{2+}, Mg^{2+}, Fe^{3+}, …) (Chen et al., 2011).

3.4 Aminoacids and small peptides as capping agents

Aminoacids and small peptides containing a thiol group like cysteine (Cys) and its derivatives and glutathione GSH (γ-Glu-Cys-Gly) have also been used to stabilize CdSe QDs in aqueous solution (Figure 23). When cadmium^{2+}-cysteine complexes (Cys/Cd^{2+} = 3) were treated by NaHSe (optimal Se^{2-}/Cd^{2+} ratio = 0.75) at pH = 10.5, CdSe@Cys QDs emitting at ca. 555 nm were obtained after 2 hours heating at 90°C. After interaction with bovine serum albumin (BSA), these QDs were used to label *E. coli* cells treated by lysozyme (Liu et al., 2009).

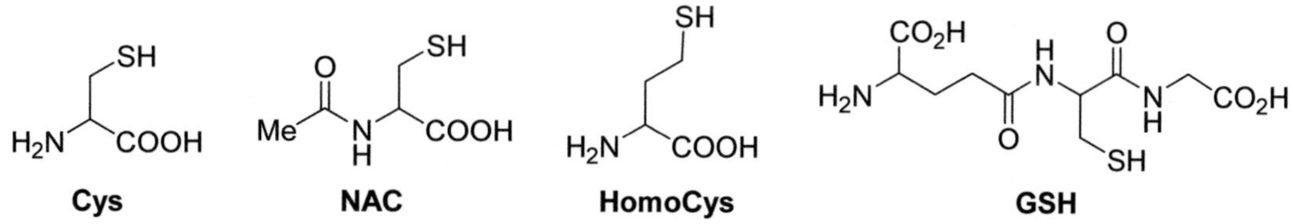

Fig. 23. Cysteine and cysteine derivatives used to stabilize CdSe nanoparticles

The size distribution and the stability of CdSe nanoparticles stabilized by cysteine or cysteine derivatives were also recently evaluated. CdSe nanocrystals were prepared at pH = 12 using a Cd/Se/Cys ratio of 4/1/8.8 at room temperature from $CdSO_4$ and Na_2SeO_3. Cysteine was found to be the best capping ligand for preparing small-sized (diameter between 1.4 and 1.8 nm) CdSe QDs with a narrow size distribution exhibiting a sharp excitonic peak at 410 nm. The nanocrystals were stable for at least 3 months in aqueous solution. The thiol group of Cys binds to the surface Cd atoms of CdSe nanoparticles. The secondary coordination was found to involve the amine group of Cys. The carboxylate group of Cys is only involved in the electrostatic stabilization of the colloidal CdSe nanoparticles and contributes to their long-term stability. The stability of CdSe nanoparticles prepared using Cys methyl or ethyl ester was maintained only for a short time confirming the crucial role played by the carboxylate group on the colloidal stability. *N*-acetylcysteine (NAC), which contains a nitrogen atom with very low nucleophilicity unable to coordinate with Cd surface atoms of the nanocrystals and thus which can only coordinate through its thiol function can be used to prepare CdSe nanoparticles but is unable to provide stability. Homocysteine, which coordinates only through the thiol function with the surface of CdSe nanoparticles because the amine group is located too far from the nanoparticles surface, can also restrict the initial growth of CdSe nanocrystals and help to maintain their short- and long-term stability (Park et al., 2010).

The same authors have recently demonstrated a strong influence of the Cys concentration on the optical properties and on the PL QYs of the CdSe QDs prepared. CdSe nanocrystals prepared at low Cys to precursor molar ratios (nCys/nCd ~ 1) exhibit intense PL QY (~ 18%) due to the presence of Cd-Cys complexes and cadmium oxides, which passivate the surface of the CdSe cores and reduce the non-radiative recombinations of excitons (Park et al., 2011).

Cysteine was also used as capping ligand for the preparation at room temperature of CdSe semiconductor magic-sized clusters (MSCs) using $CdSO_4$ and Na_2SeSO_3 as starting materials. The room temperature synthesis conditions ensured both very slow nucleation and growth, favoring focusing of the size distribution. As demonstrated by mass spectrometry, using a Cd/Se/Cys ratio of 1/0.25/8.8 and allowing the nanoparticles to growth for 7 days afforded $(CdSe)_{33}$ and $(CdSe)_{34}$ MSCs exhibiting a sharp first absorption peak at 420 nm. The sharpness of the first absorption peak indicates a very narrow size distribution. The average diameter of the nanoparticles produced was estimated to be 1.6 nm by TEM, by AFM, and by the optical absorption spectrum. The PL emission spectrum of the CdSe MSCs is composed of a sharp peak at 429 nm (excitonic emission) and of a broad longer-wavelength feature (surface trap emission) (Park et al., 2010).

Optical performances of glutathione-capped CdSe QDs were also recently studied. The dots were prepared from $CdCl_2$ and H_2Se in the presence of reduced GSH at pH = 11.5 and at 90°C and have diameters between 3 and 10 nm (Figure 24). A low concentration of GSH during the synthesis (optimal Cd^{2+}/GSH ratio = 1.42) favour the production of CdSe QDs with high PL QYs. Absorption, PL emission and PL QYs of CdSe@GSH QDs were monitored upon aging at room temperature in air and with and without illumination. PL QYs were found to increase with ageing but the increase was more marked under illumination. As demonstrated by XRD, photo-induced decomposition of the surface of the QDs under illumination generates CdO and Se which remove surface defects and improve PL QYs (the highest PL QY, 36.6%, was reached after 44 days storage) (Wang et al., 2011).

$$4\,NaBH_4 + Se + 7\,H_2O \longrightarrow 2\,NaHSe + Na_2B_4O_7 + 15\,H_2$$

$$2\,NaHSe + H_2SO_4 \longrightarrow Na_2SO_4 + 2\,H_2Se$$

$$H_2Se + CdCl_2 + 2\,NaOH \longrightarrow 2\,NaCl + CdSe + 2\,H_2O$$

Fig. 24. Synthesis of CdSe QDs from H_2Se and $CdCl_2$

4. Conclusions and outlook

CdSe nanocrystals are important II–VI semiconductor materials with high luminescence quantum yields. The organometallic synthesis, injection of reagents into a coordinating solvent such as tri-*n*-octylphosphine oxide (TOPO) at high temperature (200-400°C), is an effective route to prepare high-quality CdSe QDs. The biological applications of these kinds of QDs have been hampered by their inherently low dispersibility in water. Then, new chemical strategies have been established to solve this problem. One is to synthesize in the aqueous solution, either by heating at 100°C, by the hydrothermal method or by microwave-assisted synthesis.

This review demonstrates that direct synthesis of CdSe nanocrystals in water is a promising alternative route to organometallic reactions because this process is economical and green and that the prepared fluorescent nanoparticles have high photoluminescence efficiencies. The most important feature is that the CdSe nanoparticles did not need further surface modification to be water-dispersible and that the surface functionalization with water-soluble ligands can be performed during the synthesis. Moreover, the size and the morphology of CdSe nanoparticles can conveniently be tuned by controlling the concentration of precursors, annealing temperature, and time of reaction.

Without any doubt, aqueous-based syntheses open up unique opportunities for CdSe quantum dots and CdSe-based nanomaterials that cannot be provided by any other synthesis techniques. Because of their size-tunable emission covering the whole visible spectrum, there are many reasons to believe that the low-dimensional CdSe nanomaterials synthesized by the aqueous route will be extremely useful for photonics, optoelectronics, and bio-imaging in the coming years. It is also reasonable to believe that the hydrothermal methods presented here could be used for large-scale fabrication of other semi-conductor nanoparticles.

5. References

Aldeek, F.; Balan, L.; Lambert, J.; Schneider, R. (2008). The influence of capping thioalkyl acid on the growth and photoluminescence efficiency of CdTe and CdSe quantum dots. *Nanotechnology*, 19, 475401 (9 pp).

Aldeek, F.; Mustin, C.; Balan, L.; Medjahdi, G.; Roques-Carmes, T.; Arnoux, P.; Schneider, R. (2011). Enhanced photostability from CdSe(S)/ZnO core/shell quantum dots and their use in biolabeling. *Eur. J. Inorg. Chem.*, 794-801.

Aldeek, F.; Mustin, C.; Balan, L.; Roques-Carmes, T.; Fontaine-Aupart, M.-P.; Schneider, R. (2011). Surface-engineered quantum dots for the labelling of hydrophobic microdomains in bacterial biofilms. *Biomaterials,* 2011, 32, 5459-5470.

Alivisatos, A.P. (1996). Semiconductor clusters, nanocrystals, and quantum dots. *Science,* 271, 933-937.

Chen, M.; Gao, L. (2005). Synthesis and characterization of cadmium selenide nanorods via surfactant-assisted hydrothermal method. *J. Am. Ceram. Soc.*, 88, 1643-1646.

Chen, M.; Gao, L. (2006). From $Cd(OH)_2$ nanoflakes to CdSe nanochains: synthesis and characterization. *J. Crys. Growth*, 286, 228-234.

Chen, S.; Zhang, X.; Zhang, Q.; Hou, X.; Zhou, Q.; Yan, J.; Tan, W. (2011). CdSe quantum dots decorated by mercaptosuccinic acid as fluorescent probe for Cu^{2+}. *J. Lumin.*, 131, 947-951.

Costa-Fernandez, J.M.; Pereiro, R.; Sanz-Medel, A. (2006). The use of luminescent quantum dots for optical sensing. *Trends Anal. Chem.*, 25, 207-218.

Deng, D.-W.; Yu, J.-S.; Pan, Y. (2006). Water-soluble CdSe and CdSe/CdS nanocrystals: a greener synthetic route. *J. Colloid Interface Sci.*, 299, 225-232.

Dlamini, N.N.; Rajasekhar Pullabhotla, V.S.R.; Revaprasadu, N. (2011). Synthesis of triethanolamine (TEA) capped CdSe nanoparticles. *Mater. Lett.*, 65, 1283-1286.

Eychmüller, A.; Rogach, A.L. (2000). Chemistry and photophysics of thiol-stabilized II-VI semiconductor nanocrystals. *Pure Appl. Chem.*, 72, 179-188.

Foos, E.E.; Wilkinson, J.; Mäkinen, A.J.; Watkins, N.J.; Kafafi, Z.H.; Long, J.P. (2006). Synthesis and surface composition of CdSe nanoclusters prepared using solvent systems containing primary, secondary and tertiary amines. *Chem. Mater.*, 18, 2886-2894.

Ge, J.-P.; Li, Y.-D.; Yang, G.-Q. (2002). Mechanism of aqueous ultrasonic reaction: controlled synthesis, luminescence properties of amorphous cluster and nanocrystalline CdSe. *Chem. Commun.*, 1826-1827.

Han, H.; Di Francesco, G.; Maye, M.M. (2010). Size control and photophysical properties of quantum dots prepared via a novel tunable hydrothermal route. *J. Phys. Chem. C*, 114, 19270-19277.

Han, H.-y.; Sheng, Z.-h.; Liang, J.-g. (2006). A novel method for the preparation of water-soluble and small-size CdSe quantum dots. *Mater. Lett.*, 60, 3782-3785.

Hao, E.; Lian, T. (2000). Layer-by-layer assembly of CdSe nanoparticles based on hydrogen bonding. *Langmuir*, 16, 7879-7881.

Huang, D.; Liu, H.; Zhang, B.; Jiao, K.; Fu, X. (2009). Highly sensitive electrochemical detection of sequence-specific DANN of 35S promoter of cauliflower mosaic virus gene using CdSe quantum dots and gold nanoparticles. *Microchim. Acta*, 165, 243-248.

Kalasad, M.N.; Rabinal, M.K.; Mulimani, B.G. (2009). Ambient synthesis and characterization of high-quality CdSe quantum dots by an aqueous route. *Langmuir*, 25, 12729-12735.

Klimov, V.; Haring Bolivar P.; Kurz, H. (1996). Ultrafast carrier dynamics in semiconductor quantum dots. *Phys. Rev. B*, 53, 1463-1467.

Li, J.; Ni, Y.; Liu, J.; Hong, J. (2009). Preparation, conversion, and comparison of the photocatalytic property of $Cd(OH)_2$, CdO, CdS and CdSe. *J. Phys. Chem. Solids*, 70, 1285-1289.

Lim, J.; Jun, S.; Jane, E.; Baik, H.; Kim, H.; Cho, J. (2007). Preparation of highly luminescent nanocrystals and their application to light-emitting diodes. *Adv. Mater., 19*, 1927-1932.

Lin, Y.-W.; Hsieh, M.-M.; Liu, C.-P.; Chang, H.-T. (2005). Photoassisted synthesis of CdSe and core-shell CdSe/CdS quantum dots. *Langmuir*, 21, 728-734.

Liu, J.; Xue, D. (2010). Morphology-controlled synthesis of CdSe semiconductor through a low-temperature hydrothermal method. *Phys. Scr.*, T139, 014075 (5 pp).

Liu, P.; Wang, Q.; Li, X. (2009). Studies on CdSe/L-Cysteine quantum dots synthesized in aqueous solution for biological labelling. *J. Phys. Chem. C*, 113, 7670-7676.

Medintz, I.L.; Uyeda H.T.; Goldman E.R.; Mattoussi, H. (2005). Quantum dot bioconjugates for imaging, labelling and sensing. *Nat. Mater.*, 4, 435-446.

Nikoobakht, B.; Burda, C.; Braun, M.; Hun, M.; El-Sayed, M.A. (2002). The quenching of CdSe quantum dots photoluminescence by gold nanoparticles in solution. *Photochem. Photobiol.*, 75, 591-597.

Owen, J.S.; Park, J.; Trudeau, P.-E.; Alivisatos, A.P. (2008). Reaction chemistry and ligand exchange at cadmium-selenide nanocrystal surfaces. *J. Am. Chem. Soc.*, 130, 12279–12281.

Palaniappan, K.; Xue, C.; Arumugam, G.; Hackney, S.A.; Liu, J. (2006). Water-soluble, cyclodextrin-modified CdSe-CdS core-shell structured quantum dots. *Chem. Mater.*, 18, 1275-1280.

Park, Y.-S.; Dmytruk, A.; Dmitruk, I.; Kasuya, A.; Takeda, M.; Ohuchi, N.; Okamoto, Y.; Kaji, N.; Tokeshi, M.; Baba, Y. (2010). Size-selective growth and stabilization of small CdSe nanoparticles in aqueous solution. *ACS Nano*, 4, 121-128.

Park, Y.-S.; Dmytruk, A.; Dmitruk, I.; Kasuya, A.; Okamoto, Y.; Kaji, N.; Tokeshi, M.; Baba, Y. (2010). Aqueous phase synthesized CdSe nanoparticles with well-defined numbers of constituent atoms. *J. Phys. Chem. C*, 114, 18834-18840.

Park, Y.-S.; Okamoto, Y.; Kaji, N.; Tokeshi, M.; Baba, Y. (2011). Aquous phase-synthesized small CdSe quantum dots: adsorption layer structure and strong band-edge and surface trap emission. *J. Nanopart. Res.*, DOI 10.1007/s11051-011-0273-7.

Peng, Q.; Dong, Y.; Deng, Z.; Sun, X.; Li, Y. (2001). Low-temperature elemental-direct-reaction route to II-VI semiconductor nanocrystalline ZnSe and CdSe. *Inorg. Chem.*, 40, 3840-3841.

Peng, Z.A.; Peng, X. (2001). Formation of high-quality CdTe, CdSe, and CdS nanocrystals using CdO as precursor. *J. Am. Chem. Soc.*, 123 , 183–184.

Peng, Q.; Dong, Y.; Deng, Z.; Li, Y. (2002). Selective synthesis and characterisation of CdSe nanorods and fractal nanocrystals. *Inorg; Chem.*, 41, 5249-5254.

Qian, H.; Li, L.; Ren, J. (2005). One-step and rapid synthesis of high quality alloyed quantum dots (CdSe-CdS) in aqueous phase by microwave irradiation with controllable temperature. *Mater. Res. Bull.*, 40, 1726-1736.

Qu, L.H.; Peng, X.G. (2002). Control of photoluminescence properties of CdSe nanocrystals in growth. *J. Am. Chem. Soc., 124*, 2049-2055.

Robel, I.; Subtamanian, V.; Kuno, M.; Kamat, P.V. Quantum dot solar cells. Harvesting light energy with CdSe nanocrystals molecularly linked to mesoscopic TiO_2 films. *J. Am. Chem. Soc.* 2006, *128*, 2385-2393.

Rogach, A.L.; Katsikas, L.; Kornowski, A.; Su, D.; Eychmüller, A.; Weller, H. (1996). Synthesis and characterization of thiol-stabilized CdTe nanocrystals. *Ber. Bunsen-Ges. Phys. Chem.*, 100, 1772-1778.

Rogach, A.L.; Kornowski, A.; Gao, M.; Eychmüller, A.; Weller, H. (1999). Synthesis and characterization of a size series of extremely small thiol-stabilized CdSe nanocrystals. *J. Phys. Chem. B*, 103, 3065-3069.

Rogach, A.L.; Nagesha, D.; Ostrander, J.W.; Giersig, M.; Kotov, N.A. (2000). «Raisin bun»-type composite spheres of silica and semiconductor nanocrystals. *Chem. Mater.*, 12, 2676-2685.

Tao, L.; Pang, S.; An, Y.; Xu, H.; Wu, S. (2010). Enhanced photoelectric activity of CdSe nanostructures with mixed crystalline phases. *Phys. Scr.*, T139, 014077 (5 pp).

Vossmeyer, T.; Reck, G.; Katsikas, L.; Haupt, E.T.K.; Schulz, B.; Weller, H. (1995). A "double-diamond superlattice" built up of $Cd_{17}S_4(SCH_2CH_2OH)_{26}$ clusters. *Science*, 267, 1476-1479.

Wang, Y.; Tang, Z.; Correa-Duarte, M.A.; Liz-Marzan, L.M.; Kotov, N.A. (2003). Multicolor luminescence patterning by photoactivation of semiconductor nanoparticle films. *J. Am. Chem. Soc.*, 125, 2830-2831.

Wang, Y.; Tang, Z.; Correa-Duarte, M.A.; Pastoriza-Santos, I.; Giersig, M.; Kotov, N.A.; Liz-Marzan, L.M. (2004). Mechanism of strong luminescence photoactivation of citrate-stabilized water-soluble nanoparticles with CdSe cores. *J. Phys. Chem. B*, 108, 15461-15469.

Wang, L.-L.; Jiang, J.-S. (2011). Optical performance evolutions of reductive glutathione coated CdSe quantum dots in different environments. *J. Nanopart. Res.*, 13, 1301-1309.

Williams, J.V.; Adams, C.N.; Kotov, N.A.; Savage, P.E. (2007). Hydrothermal synthesis of CdSe nanoparticles. *Ind. Eng. Chem. Res.*, 46, 4358-4362.

Williams, J.V.; Kotov, N.A.; Savage, P.E. (2009). A rapid hot-injection method for the improved hydrothermal synthesis of CdSe nanoparticles. *Ind. Eng. Chem. Res.*, 48, 4316-4321.

Xie, Y.; Wang, W.Z.; Qian, Y.T.; Liu, X.M. (1999). Solvothermal route to nanocrystalline CdSe. *J. Solid State Chem.*, 147, 82-84.

Xue, X.; Pan, J.; Xie, H.; Wang, J.; Zhang, S. (2009). Fluorescence detection of total count of *Escherichia coli* and *Staphylococcus aureus* on water-soluble CdSe quantum dots coupled with bacteria. *Talanta*, 77, 1808-1813.

Yan, Y.-l.; Li, Y.; Qian, X.-f.; Yin, J.; Zhu, Z.-k. (2003). Preparation and characterization of CdSe nanocrystals via Na_2SO_3-assisted photochemical route. *Mater. Sci. Eng. B*, 103, 202-206.

Yang, J.; Zang, C.; Wang, G.; Xu, G.; Cheng, X. (2010). Synthesis of CdSe microspheres via a solvothermal process in a mixed solution. *J. Alloys Compd.*, 495, 158-161.

Yang, Q.; Tang, K.; Wang, F.; Wang, C.; Qian, Y. (2003). A γ–irradiation reduction route to nanocrystalline CdE (E = Se, Te) at room temperature. *Mater. Lett.*, 57, 3508-3512.

Yu, S.-H.; Wu, Y.-S.; Yang, J.; Han, Z.-H.; Xie, Y.; Qian, Y.-T.; Liu, X.-M. (1998). A novel solvothermal synthetic route to nanocrystalline CdE (E = S, Se, Te) and morphological control. *Chem. Mater.*, 10, 2309-2312.

Zhong, P.; Yu, Y.; Wu, J.; Lai, Y.; Chen, B.; Long, Z.; Liang, C. (2006). Preparation and application of functionalized nanoparticles of CdSe capped with 11-mercaptoundecanoic acid as a fluorescent probe. *Talanta*, 70, 902-906.

Zhu, J.-J.; Xu, S.; Wang, H.; Zhu, J.-M.; Chen, H.-Y. (2003). Sonochemical synthesis of CdSe hollow spherical assemblies via an in-situ template route. *Adv. Mater.*, 15, 156-159.

9

Stimulated Formation of InGaN Quantum Dots

A.F. Tsatsulnikov and W.V. Lundin
Ioffe Institute
Russia

1. Introduction

Interest to the formation of QDs in wide bandgap InGaAlN system is due to several reasons. First, all layers in an InGaAlN-based heterostructures are lattice mismatched that leads to strong phase separation and QD formation even in low content thin InGaN layers. Second, QDs in the wide bandgap material have large localization energy that leads to effective localization of carriers, suppression of carrier transport toward defect areas, density of which typically is about 5-10×108 cm-3, and resulting in effective emission efficiency. Third, InGaN QDs allow expanding of emission range of InGaN-based light emitting devices. For InGaN-based light emitting diodes (LEDs) even red emission was demonstrated (Mukai et al., 1999). However, spontaneous formation of the InGaN QDs requires development of the special methods to control their structural and optical properties. In this sense detail investigations of the correlation between parameters of growth of the QDs and their properties (density, sizes, content) are important. Different approaches were studied to form InGaN QDs. Stranskii-Krastanov mechanism of InGaN QD formation was investigated in (Bai et al., 2009; Choi et al., 2007). However, specific growth conditions are required to form such QDs that can leads to worsening of structural quality of the InGaN layers and low efficiency of emission. More attention was paid to investigations of the phase separation in thin InGaN QWs at changing of the equilibrium of the epitaxial growth conditions. It was show that different growth parameters affect the properties of the InGaN QDs: annealing of the InGaN QWs after deposition (Oliver et al., 2003); growth conditions of GaN layer covering the InGaN QW (Wang et al., 2008; Wen et al., 2001); growth interruptions (GIs) after deposition of the InGaN QWs (Choi et al., 2007; Ji et al., 2004); changing of In flow or growth temperature during InGaN deposition (Kumar et al., 2008; Musikhin et al., 2002; Shim et al., 2002 ; Soh et al., 2008 ; Sun et al., 2004). Thus, formation of the InGaN QDs are depending on large number of parameters. For complex multilayer structure changing of structure design and growth conditions of under layers (below active region) can leads to changing of the active region properties. Moreover, this is correct for the effect of upper layers (above active region). This can be critical, for example, in the case of growth of monolithic white LEDs containing several InGaN QWs emitting at different wavelengths in wide region of spectrum. Besides, effect of QDs on width of emission line can improve color parameters of RGB, in particular monolithic, white LEDs.

In this paper we tried to paid attention to the results of investigations which revealed effect of different growth parameters on QD formation which usually are not widely discussed (e.g. growth pressure, growth atmosphere, layer sequence etc.) and considered features in an LED properties which are associated with QD formation. Effect of the QD formation on such important parameters of LEDs as wavelength and current dependencies of the emission wavelength and quantum efficiency will be shown.

2. Technological aspects

Epitaxial structures described below were grown by Metalorganic Vapor Phase Epitaxy (MOVPE) using systems with horizontal flow (strongly re-designed Epiquip VP-50RP) and Planetary (AIX2000HT) reactors. Gas blending units of the systems have allowed us to use hydrogen, nitrogen or their mixture at any given ratio as a carrier gas. Ammonia, trimethylgallium (TMGa), triethylgallium (TEGa), trimethylindium (TMIn), trimethylaluminum (TMAl), biscyclopentadienylmagnesium (Cp_2Mg), and silane (SiH_4) were used as precursors. Reactors were equipped with a home-made *in-situ* optical reflectance monitoring (ORM) tools. Most of epitaxial structures were grown on (0001) sapphire substrates utilizing conventional low-temperature GaN nucleation technique. Experimental structures were characterized by various structural, electrical, and optical methods including scanning (SEM) and transmission (TEM) electron microscopy, atomic force microscopy (AFM), secondary ion mass spectrometry (SIMS), X-ray diffractometry, Hall-effect measurements, and spectroscopy of photo (PL) and electroluminescence (EL). Thickness, alloy composition and homogeneity of individual InGaN QWs, InGaN and GaN layers were investigated by Digital Analysis of Lattice Image (DALI) (Gerthsen et al., 2000; Rosenauer et al., 1996) or Geometrical Phase Analysis (GPA) (Hytch et al., 1998) of High Resolution transmission electron microscopy (HRTEM) images.

3. Methods of QD formation

In this chapter different methods of InGaN QD formation will be considered. We will focus on an in-situ methods of QD formation i.e. methods when QDs are formed directly during growth of InGaN QWs. These methods include different technological approaches when QD formation is stimulated by certain growth conditions, for example, growth temperature, GIs, atmosphere and pressure. On the other hand, QD formation can be stimulated by applying of special design (layer sequence) of structures.

3.1 Effect of growth temperature

Difference in optimal growth conditions for InGaN and GaN requires fast change of the reactor conditions when switching between these materials. In particular, GaN barriers are typically grown at temperatures at least 50-70°C higher than InGaN. This temperature ramp usually takes more than 20-30 sec. Thus one has to choose between growth interruption (GI) and growth at ramped conditions. Both alternatives have disadvantages. Growth during reactor conditions change generally is less reproducible. Moreover, only part of GaN barrier is grown at higher temperature. On the other hand, during GI InGaN surface is exposed to the reactor atmosphere for relatively long time that influences structure properties. There are a number of publications concerning GI-related effects (Cho et al., 2001; Daele et al., 2004). All authors describe blue shift and indium excess removal as a result of GI but general conclusions on the GI benefit is contradictory: blue shift sometimes is too high price for quality improvement. Moreover, depending on growth parameters, GI can either increase In nonuniformity in the InGaN layer or make this layer more uniform. Thus, effects of temperature and GIs on InGaN properties should be investigated consequently.

In the work (Musikhin et al., 2002) InGaN/GaN MQW structures were grown by applying of the method of thermocycling, when the MQW structures are formed by cycled variation of growth temperature (in temperature range between T_{min} and T_{max}) at constant gas flows that realize periodical changes of the In content. Fig. 1 (a, b) shows high resolution TEM (HRTEM) images of these structures grown in temperature range of 755-870°C and 785-900°C. It is seen that distribution of the In atoms in the samples grown by this method is inhomogeneous with clearly resolved In-rich QDs. The decrease of the growth temperature results in an increase in the In concentration with average values of 18% for sample grown at high temperature and 23% for sample grown at low temperature. The total amount of InN incorporated into the QW was calculated on the basis of the maximum displacement at the top of InGaN layer yielding 1.5 monolayers (ML) and 2 ML for samples grown at higher and lower temperatures, respectively. PL spectra of the sample grown at a lower temperature show a 200 meV shift of the emission peak to the long wavelength side that agreed with increase in the In content.

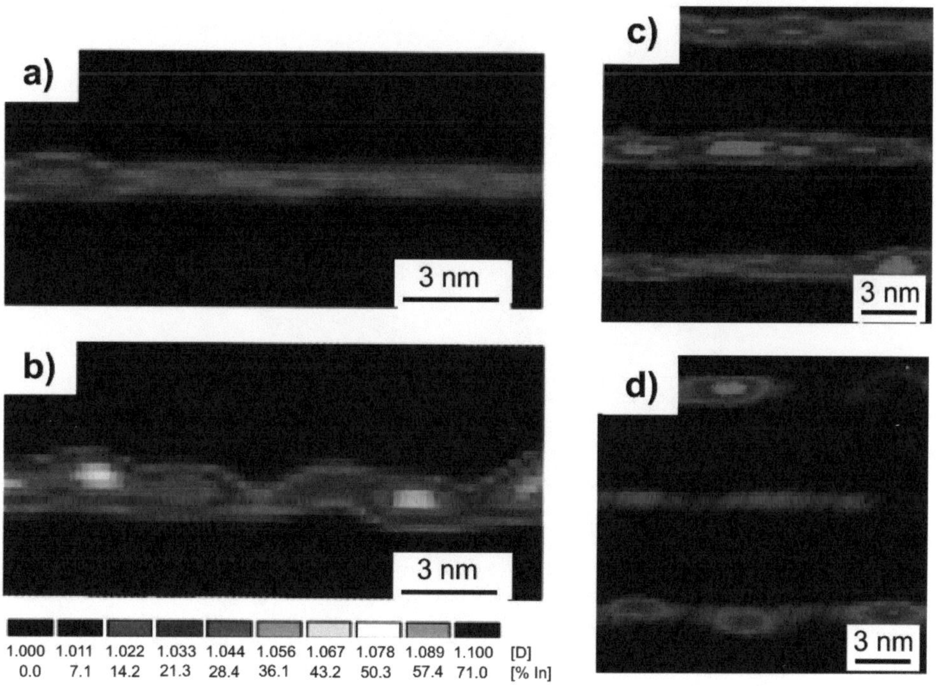

Fig. 1. Color-coded maps of the local In concentration obtained by digital analysis of high-resolution transmission electron microscopy (HRTEM) images of the samples grown at lower (a) and higher (b) substrate temperatures; at higher (c) and lower (d) TMI/TMG-flow ratios. (Musikhin et al., 2002)

HRTEM shows that lateral size of the QDs are about 3 nm for the samples lower growth temperature (Fig. 1b). Decreasing the InGaN growth temperature (from 785 °C to 755 °C) results in: a slight increase in QD size; increase of the maximum In concentration in the

QDs (from 35% to 50% for the growth temperature change), and increase of the QD density (from 0.5×10^{12} cm^{-2} to 3×10^{12} cm^{-2} for the growth temperature change). Thus, relative change in the growth temperature of the InGaN/GaN QWs at some TMIn flow leads to enhance of the effect of phase separation. Absolute In content at such growth method may be also changed by TMIn/TMGa ratio. For the investigated regimes it even stronger influence indium content than growth temperature (fig 1) which is in a good agreement with results of other works (Bedair et al., 1997; Schenk et al., 1999; Yoshimoto et al., 1991).

The key experiment, which can clarify unambiguously the internal nature of the InGaN QW grown at these conditions, is the resonant excitation of PL (Krestnikov et al., 2002a; Krestnikov et al., 2002b).

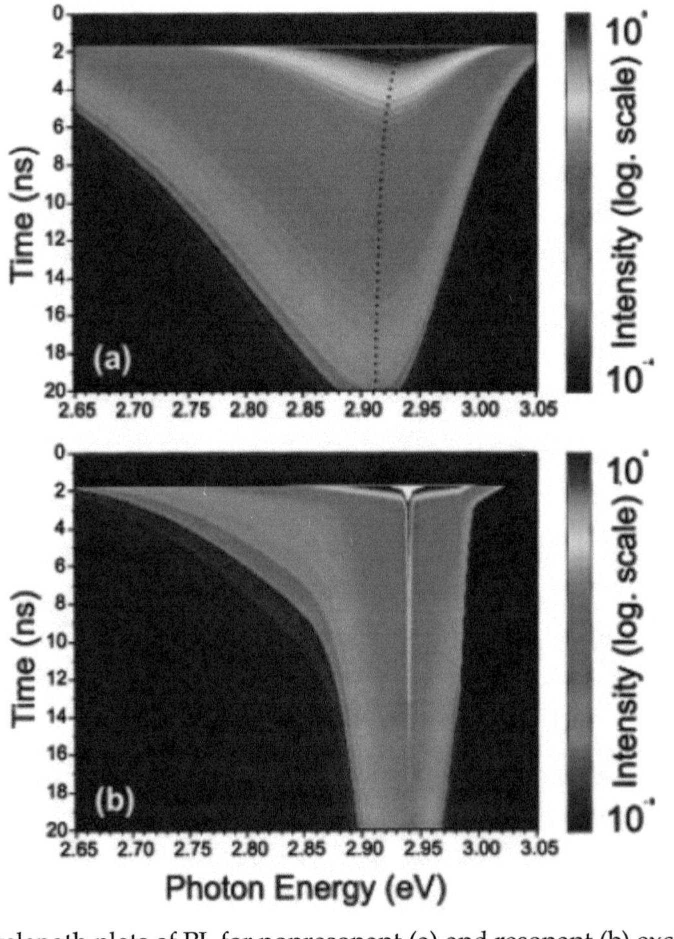

Fig. 2. Time/wavelength plots of PL for nonresonant (a) and resonant (b) excitations for sample grown at 820 °C. White spot corresponds to the scattered laser light. (Krestnikov et al., 2002a)

In the case when the QW is excited, the carrier transfer towards the states having a lower energy occurs, accompanied by fast evolution of the thermalized PL peak. In the case of wurtzite-type strained QWs having a strong built-in piezoelectric field, reduced carrier concentration upon the radiative recombination of nonequilibrium carriers must result in reduced screening of the piezoelectric potential, further shifting PL towards the longer-wavelength side of the spectra and strongly increasing the PL decay time. Neither of these effects was revealed in the considered case. Nonresonant pulsed excitation results in a broad PL peak, while resonant excitation into the nonresonant PL intensity maximum results in an evolution of a sharp resonant PL peak, having a spectral shape defined by the excitation laser pulse and a radiative decay time close to that revealed for PL under nonresonant excitation. The PL peak position was not shifted with time and no spectral broadening was observed (Fig. 2). This may rule out any importance of the spectral diffusion and gradual weakening of the piezoelectric screening with carrier depopulation, which both should take place in the case of QW. The observed behavior fits exactly the behavior of resonantly excited QDs having a delta-function-like density of state (Paillard et al., 2000) and no exciton or carrier transfer at low temperatures. Observation of a resonantly excited narrow PL line gives clear proof of the QD nature of luminescence in InGaN-GaN samples growth using the described above method.

3.2 GaN and InGaN interaction with hydrogen

Before presenting more complex methods of QDs formation we should briefly describe the effect of III-N materials interaction with hydrogen. This effect looks to be the main difference between MOVPE of III-nitrides and classical III-V compounds (Lundin et al., 2005, Yakovlev et al., 2008). The most known manifestation of hydrogen influence on III-N epitaxial process is an insistence of InGaN growth in hydrogen-free ambient (Nakamura et al., 1993) due to suppression of indium incorporation into InGaN layers by hydrogen (Piner et al., 1997). At the same time, hydrogen influences growth and properties of practically all layers in device structures starting from nucleation layer and up to p-contact layer (Lundin et al., 2009a; Lundin et al., 2009b; Yakovlev et al., 2008).

Under typical MOVPE conditions with hydrogen as a carrier gas, GaN epilayer etching occurs, if TMGa supply is switched off (Lundin et al., 2005; Sakharov et al., 2000; Zavarin et al., 2005). It was observed that under any useful reactor conditions GaN epilayer morphology during etching was kept planar allowing subsequent epigrowth. It was proved by a special investigation that GaN growth rate was not influenced by a previous etching. Thus, during one experiment growth and etching rate was measured for a number of reactor conditions (See fig. 1). The detailed study of the effect was a base for development of a kinetic model of hydrogen interaction with the surface of III-nitride layers originally suggested in (Zavarin et al., 2005) and further developed in (Yakovlev et al., 2008).

In accordance with the developed model, GaN interaction with hydrogen during MOVPE is described by the combination of two coupled processes, including reversible decomposition of GaN by H_2 with the formation of free ammonia and adsorbed gallium and reversible evaporation of the adsorbed gallium. The process of gallium desorption that determines the etching rate is strongly temperature-dependent and thus considerable only at high temperatures. GaN decomposition is a fast quasi-equilibrium process of "crystal bulk-adsorption layer" interaction which occurs in a wide temperature range and is mostly responsible for the surface coverage with gallium. This two coupled processes influence on III-N epitaxy in different ways. Gallium (and indium) evaporation from the adsorbed layer influence growth rate, alloy composition, peculiarities of initial nucleation during heteroepitaxy. Presence of this adsorbed layer with high atom mobility determines mass-transport phenomena and degree of equilibrium of material synthesis. Using of these effects in for control of III-N epitaxial process was described in (Lundin et al., 2009b; Lundin et al., 2011a). Below we will concentrate on InGaN/GaN structures fabrication.

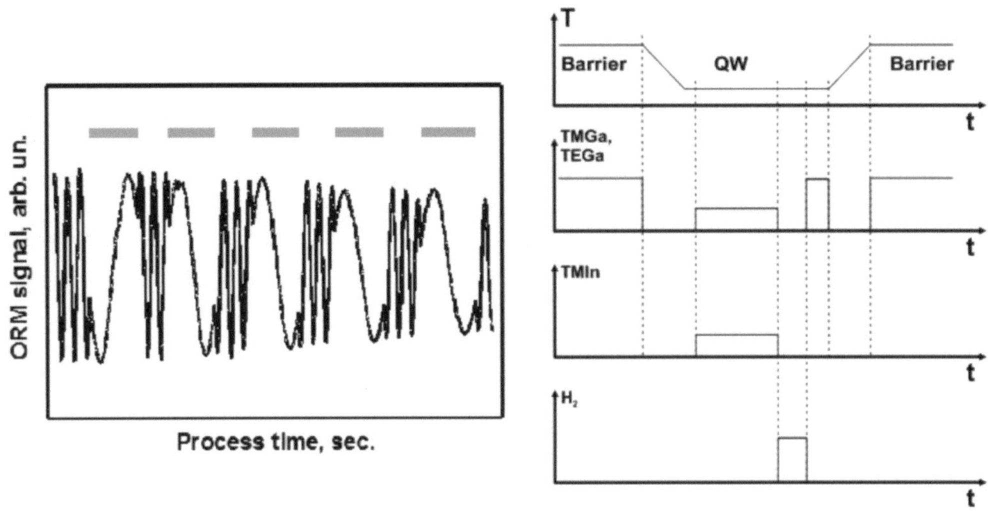

Fig. 3. a - a fragment of as-recorded ORM signal *vs* time. b - periods of etching are marked. Flow sequence for one period of InGaN/GaN MQW growth using growth interruptions with hydrogen admixing. (Lundin et al., 2009b)

3.3 Effect of hydrogen on InGaN QW formation

As it was mentioned above, using of GI after InGaN QW growth allows to improve luminescence efficiency but blue shift due to indium evaporation during GI sometimes is too high price for quality improvement. In (Sakharov et al., 2000) a modification of simple GI procedure was proposed. It was shown that GI with admixing of a small amount of hydrogen (GI+H2) into the reactor atmosphere results in significantly stronger improvement of material quality (luminescence efficiency) than in the case of GI without hydrogen for the same blue-shift.

Further development of this technology has led to separation of GI+H2 and temperature ramp (Fig. 3). Optimized procedure includes InGaN QW growth, GI+H2 and 2 nm GaN cap growth at the same temperature, GI without hydrogen for temperature ramp, and growth of GaN barrier. For MQW formation this procedure should be repeated. Maintaining of the low temperature during GI+H2 reduces blue shift keeping all benefits of InGaN exposition to hydrogen, while during consequent temperature ramp InGaN is already protected by GaN cap layer. It should be noted that the optimal hydrogen content in a carrier gas during GI+H_2 strongly depends not only on other reactor conditions, but also on the reactor design. For example, in a similar blue-LED process for small horizontal and much larger Planetary reactors used it is 0.5-1% and 20-25%, respectively.

Detailed mechanism of InGaN - hydrogen interaction during GI+H2 is unclear yet. In-N bonds are weaker than Ga-N thus etching is faster for InN component of InGaN alloy. That is why InGaN should be grown without hydrogen. During GI+H2 high-quality InGaN decomposes slower than low-quality one. This is a dominant effect for moderate quality structures and it was a driving force for our first experiments on GI+H2. In the works (Liu et al., 2003; Moon et al., 2001) it was shown that the admixing of hydrogen during GIs leads to elimination of the excess of In atoms with InGaN surface and improve structural quality of an InGaN.

Besides, there is a higher-order effect, which becomes dominant for high-quality structures. InGaN interaction with H_2 results in the increase of metallic surface coverage at the expense of atoms released from top few nanometers of crystal. Metallic indium partially evaporates, migrates over the surface, and probably re-incorporates back into the crystal volume. Similar process occurs with gallium except the gallium evaporation is low at InGaN growth temperature. Since the InGaN layer grows with the formation of a dense array of the In-rich QDs, which can be partially or completely relaxed in the nonovergrown state, the In atoms migrate to these QDs during GI, which is caused by a decrease of the elastic stresses in them. This assumption agree with the results of work (Karpov et al., 2004), where it was shown that the studied compositions correspond to the immiscibility region in In with considerable phase separation. Thus, a summary effect of these processes (evaporation and migration of In and Ga atoms) leads to a decrease in the sizes of QDs or their complete disappearance. Below there are some more results concerning influence of hydrogen treatment of InGaN QWs on their properties.

In the work (Yong-Tae Moon et al., 2001) it was shown also that the GIs+H2 decrease In content in the upper part of InGaN layer that results in short wavelength shift of the emission. However, in the case of growth thin InGaN QWs effect of the GIs is more complex and leads to modification of microstructure of whole InGaN QW (Tsatsulnikov et al., 2011a). HRTEM images of the thin InGaN layers grown with different duration of the GIs after deposition of the InGaN are shown in the Fig. 4 (a-c).

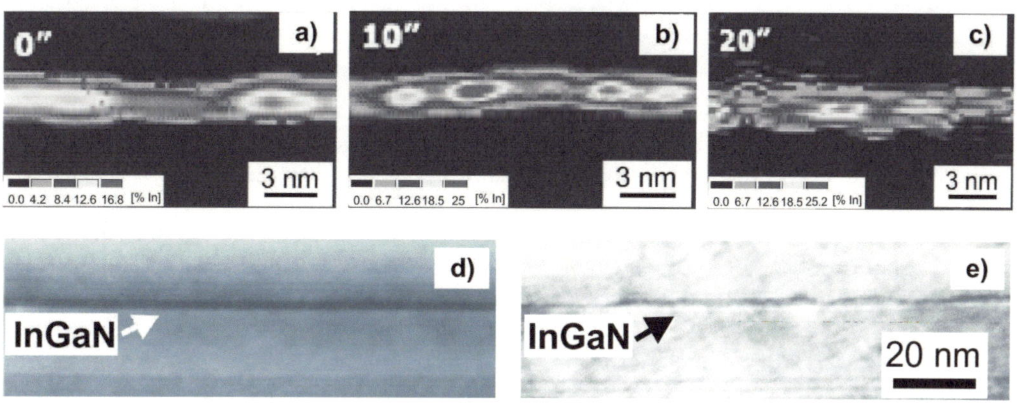

Fig. 4. Color-coded maps of the local In concentration obtained by digital analysis of HRTEM images of the samples grown with GIs having different duration in the hydrogen containing atmosphere (a-c). HRTEM images of the InGaN QWs grown with (d) and without GI (e). (Tsatsulnikov et al., 2011a)

GIs in the hydrogen containing atmosphere lead to the following changes in the structural properties of the InGaN layers: decrease in the total amount of In atoms in the layer, decrease in sizes of the formed In-rich QDs. A decrease in the total amount of In in the layer and a decrease in the sizes of the QDs is caused by conversion of the InGaN layer in GaN during GI. The joint effect of the decrease in the sizes of QDs and total amount of In lead to the 40-60 meV shift of the emission line to the higher photon energies. Except effect on the properties of the In-rich QDs the GIs resulted in transformation of the continuous InGaN layer to large isolated islands having sizes on several tens of nanometers (Fig. 4 d, e).

Another effect was observed in the case of admixing of hydrogen directly during growth of the InGaN QWs. Distribution of the In atoms in InGaN QWs grown in an atmosphere of nitrogen and with the admixing of hydrogen are shown in Figs. 5. Admixing of hydrogen during the growth leads to a considerable decrease in the density of local In-rich QDs. Comparison of the obtained results with HRTEM images obtained for the structures grown with GIs in the hydrogen containing atmosphere shows that GIs lead to disappearance of already formed QDs while the admixing of hydrogen during the growth of the InGaN initially suppresses the formation of such QDs.

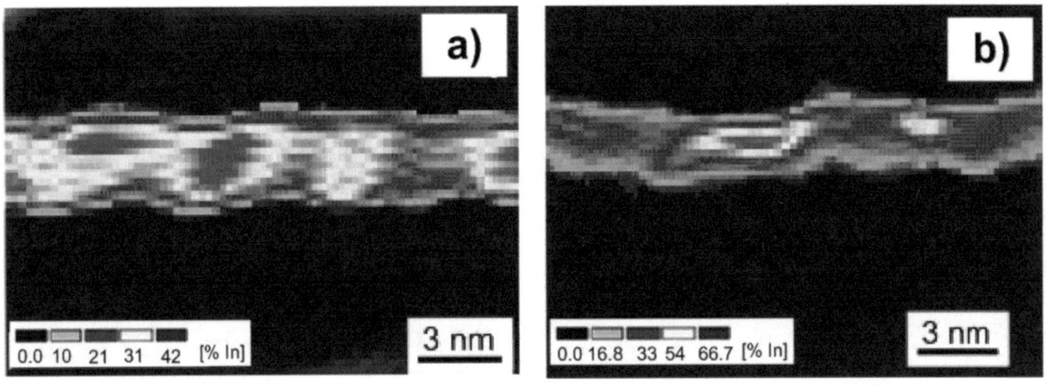

Fig. 5. HRTEM images of the InGaN QWs grown in nitrogen atmosphere (a) and with admixing of hydrogen during growth (b). (Tsatsulnikov et al., 2011a)

The studies of PL spectra upon applying the reverse bias (Tsatsulnikov et al., 2011a) show different behavior of the PL of the structures grown with and without GIs (Fig. 6). Spectrum of the structure grown without GIs has two lines. Line 2 has strong dependence on external bias. Applying the external reverse bias leads to an increase in the band bending, shift of the emission line 2 to shorter wavelengths, and drop in its intensity that indicates that this line is associated with recombination in continuous spectrum. Line 1 does not depended on the bias and can be attributed to the recombination in the local In-rich QDs. Only one line is observed in the PL spectrum of sample grown with GIs. The absence of the shift of this line and a small variation in its intensity at applying of the reverse bias indicate that recombination in this structure proceeds through localized states of QDs in which Stark effect is suppressed.

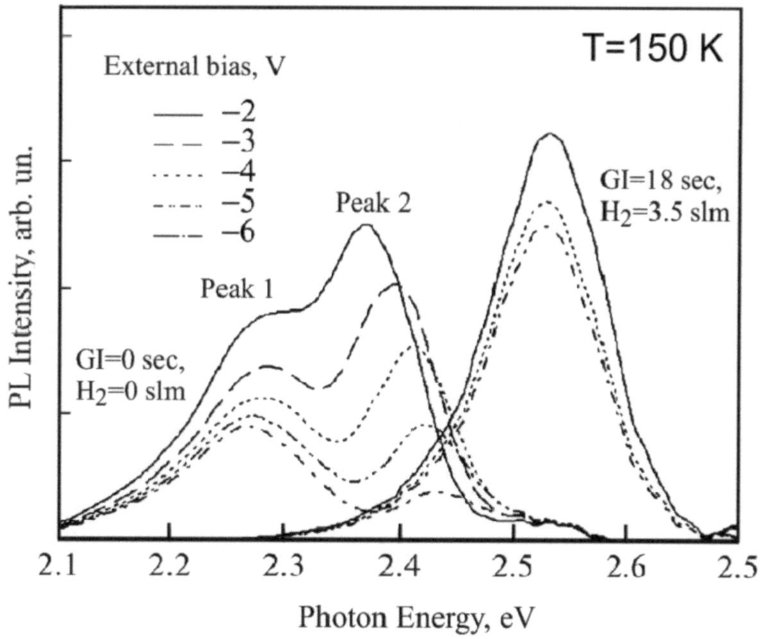

Fig. 6. Photoluminescence spectra of the InGaN/GaN QWs at various values of the reverse bias. (Tsatsulnikov et al., 2011a)

Thus, common investigations show that GI+H2 strongly affect optical and structural properties of the InGaN/GaN QWs.

3.4 Effect of reactor pressure

Fig. 7 a, b show HRTEM image of the InGaN MQW structure in which bottom InGaN QW was grown at 100 mbar and following four InGaN QWs were grown at 600 mbar (Lundin et al., 2010). It is seen that continuous InGaN layer is formed at 100 mbar. Increase of the pressure to 600 mbar resulted in formation of the separate QDs. Comparison of the emission properties of the blue LED structures with active region based on the InGaN MQWs grown at pressures in the range of 100-900 mbar shows that increase in the pressure leads to the short wavelength shift of the emission. When comparing structures emitting at different wavelengths, it is more proper to represent the value of the shift in energy units and normalize it to the difference between the energies of emitted photons and the band gap of the barrier layer (GaN). The dependences are shown in Fig. 7 c. It can be seen that, as pressure is increased, the shift of emission to shorter wavelengths becomes larger as the current increases. This effect indicates formation of individual InGaN QDs, which have a significant dispersion on sizes and content which is due to broadening of the spectrum of state. Consequently, as the current is increased, occupancy of QD states with a lower localization energy increases, which leads to an increase in the width of the emission line (Fig. 7 d). In addition, an increase in the pressure leads to variation in the relative value of the shift of the maximum ("blue shift") of the emission wavelength as a function of current (Fig. 7 e). This shift is also related to fluctuations in the In content in the InGaN layers (Martin et al., 1999).

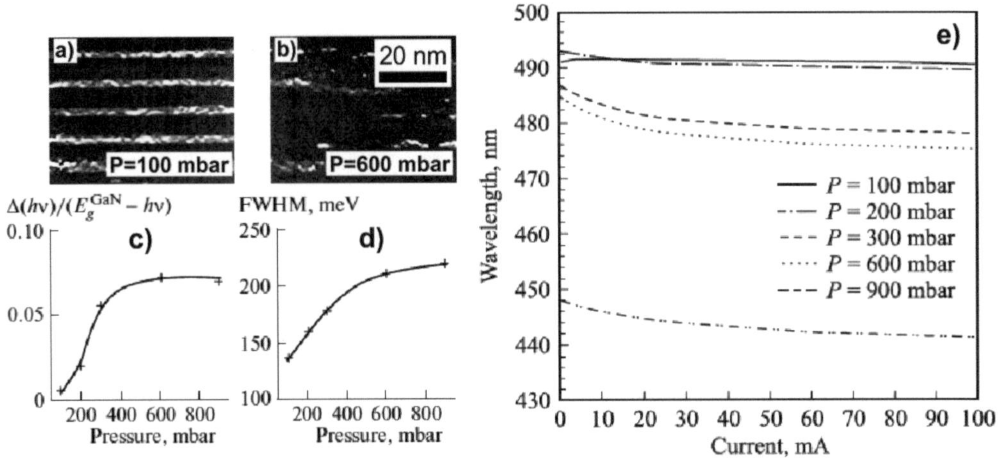

Fig. 7. HRTEM image (a, b), pressure dependencies of the relative shift of the maximum (b) and FWHM (c) of the emission and current dependencies of the maximum of the emission (d) of the InGaN QWs grown at different pressures. (Lundin et al., 2010)

Thus, in the studied series of structures, we observe an anomalous interrelation between the emission wavelength and the line width: structures with a larger wavelength show a narrower emission spectrum and a smaller shift of the emission maximum with current. An increase in the EL wavelength for InGaN/GaN structures is attained by decreasing the growth temperature (which is the generally accepted method of controlling the emission wavelength), whereas structures with a larger wavelength feature a wider spectrum and a larger shift of the emission maximum with current (Martin et al., 1999).

On the other hand, dependence of the external quantum efficiency (EQE) on the pressure has complex character. Increase in the pressure from 100 mbar to 300 mbar leads to increase of the EQE. Following increasing of the pressure to 900 mbar is resulted in decreasing of the EQE. The increase of the EQE, indeed, is due to formation of the separate InGaN QDs that suppress transport of the carriers inside InGaN QW toward defect regions. Future decrease of the EQE is due to decreasing of the carrier confinement in the InGaN QWs due to enhance of the effect of island formation.

3.5 Composite InAlN/GaN/InGaN QDs

Besides QD formation via applying of specific growth regimes, QDs can be formed in specially designed structures where this formation is forced by sequence of layers. Novel method of the QD formation based on the growth of composite InAlN/GaN/InGaN QDs was proposed in (Tsatsul'nikov et al., 2010a). This method is based on the formation of the dense array of 3D stressors from wide band gap material InAlN deposited on GaN surface (Fig. 8 a) with following deposition of the narrow band gap InGaN layer. In these structures transformation of the InGaN layer to array of isolated QDs was observed by HRTEM (Fig. 8 b). HRTEM image of the sample with 9 nm InAlN, 6 nm GaN and 3 nm InGaN layers demonstrates complex structure of the InAlN stressors having strong phase separation. This structure is composed of three regions with different morphologies and In and Al distributions in the growth direction, designated as 1, 2, and 3 in Fig. 8 b. Layer 1 is a 4 nm thick continuous 2D $In_{0.02}Al_{0.98}N$ layer with rather abrupt boundaries and constant composition across the layer. Layer 2 is also 2D, but with a nonuniform distribution of In (and Al) across its thickness. It comprises two parts. The first part begins at the boundary with the lower $In_{0.02}Al_{0.98}N$ layer and has a thickness of 2.5 nm. The concentration of In atoms in this part increases in the growth direction from 2 to 17%. The second part of the InAlN layer is 5.5 nm thick. At the boundary between the first and second parts, the content of In atoms decreases to 10% and then increases in the growth direction to 22%. Layer 3 is formed on the surface of layer 2. It is an array of 3D islands having the form of truncated pyramids with planar bases and lateral faces with base sizes of 20-30 nm and heights of 4-5 nm. The islands are closely adjacent, without any visible space between the pyramid bases, in agreement with the AFM data.

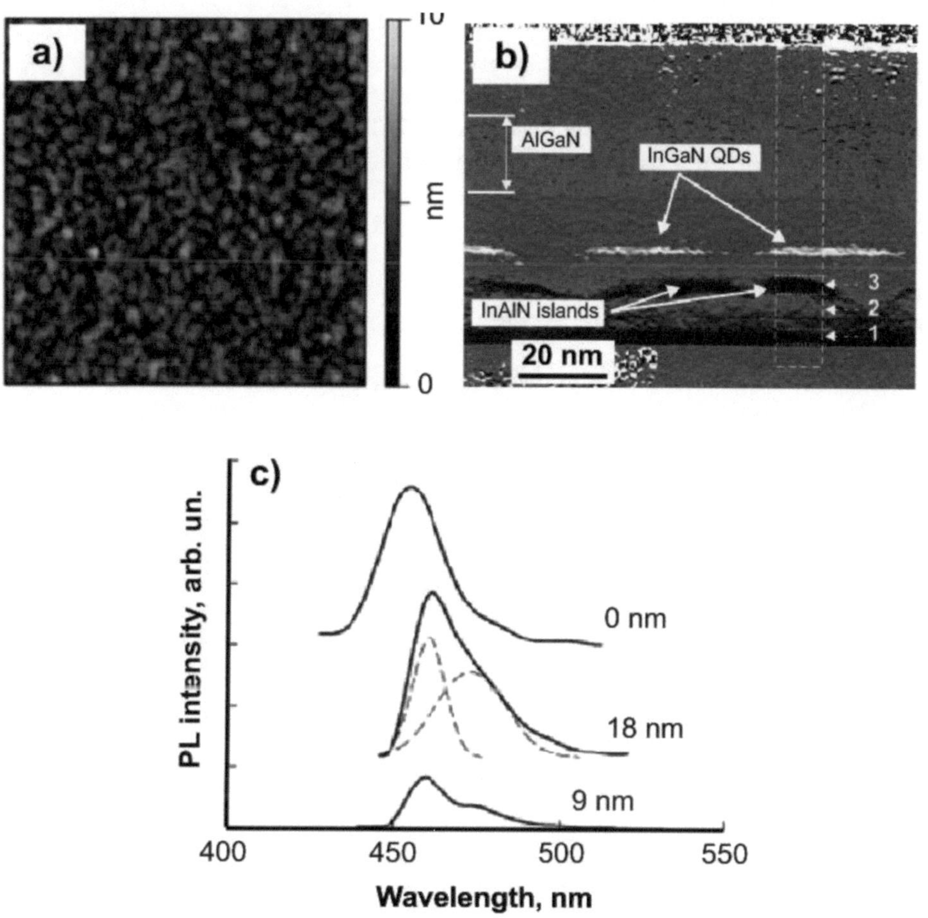

Fig. 8. AFM image (a) of the 9nm thick InAlN stressors, GPA of HRTEM (b) and PL spectra (c) of the InAlN (9 nm) /GaN (6 nm) /InGaN (3 nm) QDs. (Tsatsul'nikov et al., 2010a)

The In content in InAlN islands varies within the range 2-5% and is constant across the island thickness. Overgrowth of layer 3 with a thin GaN layer, followed by deposition of an InGaN layer, results in formation of the InGaN layer as an array of isolated InGaN islands with an average lateral size of ~30 nm, height of 2.2 nm, and average distance between the islands of ~10 nm. The average In content in the islands is 20-22%. Formation of the composite QDs leads to long wavelength shift of the emission and change in the shape of the PL line (Fig. 8 c).

3.6 Formation of nanocomposite InGaN/GaN short-period superlattices

InGaN/GaN short-period superlattices (SPSLs) are an important part of various QD-based LED structures described below. In general, InGaN/GaN SPSL may be grown by consequent growth of InGaN and GaN layers. However, an alternative approach based on InGaN interaction with hydrogen during GI may be used. If after growth of InGaN layer GI with hydrogen admixing is carried out, the described above process results in conversion of InGaN layers at the surface into GaN. In (Kryzhanovskaya et al., 2010; Tsatsulnikov et al., 2011b) InGaN/GaN SLs were formed by altering of growth of thin layers of $In_{0.1}Ga_{0.9}N$ alloy and growth interruptions when TEGa and TMIn were switched off and carrier gas composition was changed from pure N_2 to mixture of N_2: H_2 = 7:3. Thickness of GaN layers formed by this method was measured by GPA of HRTEM images with accuracy about 1 nm. It was revealed that the thickness of GaN layers formed during GI +H2 is about 1 nm for 10-20 seconds GI duration and saturates at about 2 nm at 80-160 seconds GI. It means that InGaN conversion process is self-terminated and thickness of GaN layers in these InGaN/GaN SL is limited to about 2 nm. In [41, 42] thickness of each $In_{0.1}Ga_{0.9}N$ layer grown between growth interruptions was 2 nm and duration of GI was 20 seconds which results in SPSL formation confirmed by XRD and HRTEM studies (Fig. 9 a).

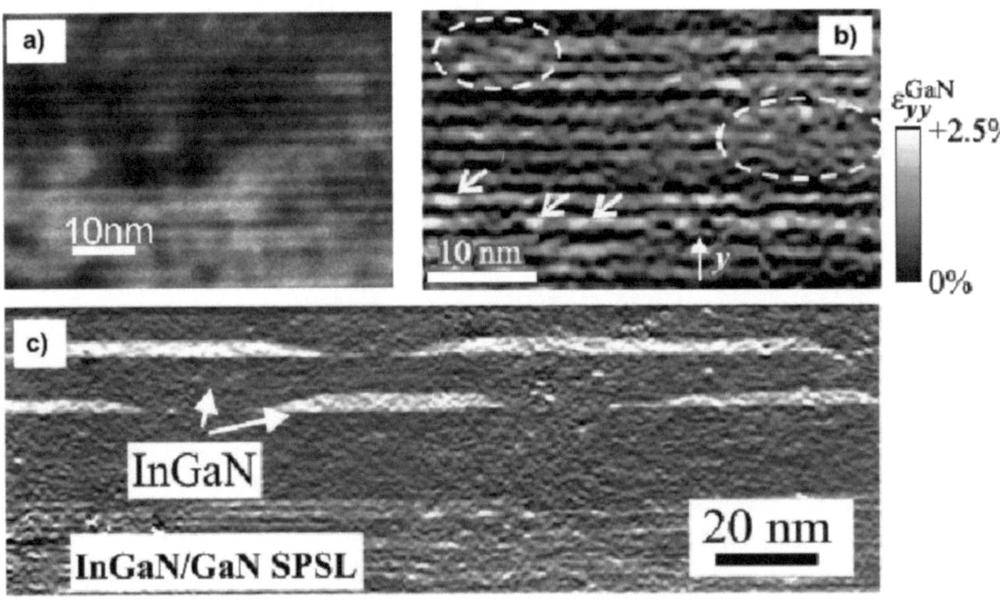

Fig. 9. TEM image (a), GPA of HRTEM (b) of the 1 nm InGaN/1 nm GaN SPSL. c - HRTEM image of the InGaN/GaN layers deposited on the InGaN/GaN SPSL covered by low temperature GaN barrier. (Kryzhanovskaya et al., 2010; Lundin et al., 2011a)

SPSL period is practically equal to the thickness of individual InGaN layer grown between GIs, and InN mole fraction periodically changes from In content in the InGaN layer to nearly zero. On the other hand, geometric phase analysis (GPA) of HRTEM images of these structures has revealed strong non-uniformity in the lateral direction at the nanometer scale (Fig. 6 b). Besides formation of In-rich local areas (arrows on the Fig. 9 b) coalescence of a neighboring InGaN layers is observed (dashed lines on the Fig. 9 b). So, these structures are not classical SPSL but can be considered as layered nanocomposite. It was revealed that deposition of the InGaN layers on the InGaN/GaN SPSL covered by low temperature GaN barrier leads to formation of large separated InGaN QDs (Fig 9 c). These features lead to complex shape of the PL spectrum of the SPSL [41].

TEM investigations of the complex structures containing the described above SPSLs have also shown that no additional dislocations were generated in SPSL in spite of relatively high thickness (up to 120 nm). Besides simple periodic SPSLs, this method was applied to the formation of periodical structures with gradually varied parameters as described below.

It should be noted that similar approach was used in (Takeuchi et al., 2009) for AlGaN/AlN SPSL growth but in this case gallium evaporation was induced by ammonia switching off.

3.7 LED structures based on the InGaN QDs

Technologies of the growth of InGaN QDs open ways for growth of the LED structures emitting in wide optical range from blue to deep green and, therefore, growth of monolithic white LED structures, containing in active region several InGaN QWs emitting in different optical ranges. QDs make wider optical range of emission due to local increase of the In content, provide better overlapping of electron and hole wave function even in the case of strong build-in piezoelectric fields, suppress carrier transport toward defect regions that leads to increasing of the emission efficiency. On the other hand, InGaN/GaN SPSL can improve injection properties of a LED active region due to effective vertical transport of carriers (Tsatsulnikov et al., 2010b). Thus, combination of the InGaN QDs and InGaN/GaN SPSL in active region allows controlling of the electrical and optical properties of a LED structures. Blue LEDs with active region of LEDs based on the InGaN MQWs confined by InGaN/GaN gradual SPSLs from both sides are described in (Tsatsulnikov et al., 2010b) (Fig. 10 a, c). Variation of indium content in SPSLs along the growth direction was realized by ramping the reactor temperature during SPSL formation procedure. An EQE of these LEDs processed in simple flip-chip geometry reached 30%.

In (Lundin et al., 2011b) deep-green LEDs were reported with an EQE of (8-20)% in the (560-530) nm range were reported. The combination of InGaN/GaN superlattice followed by low temperature GaN (Fig. 10 b, d) was shown to be the key element to increase the electroluminescence efficiency for deep-green range. It was no strain relaxation caused by SPSL in these structures in contrast to reported in (Sakharov et al., 2009), where deep-green emission was realized due to increase of indium incorporation resulting from strain relaxation in InGaN QWs below the active well responsible for light emission.

As was described above applying of the GIs after deposition of the InGaN QWs leads to the QD formation and, therefore, to features in the LED properties. GIs in the hydrogen containing atmosphere effect on the position of the emission line and current dependence of the external quantum efficiency (EQE(I)) of the LED. Applying of the GIs results in short wavelength shift of the emission (Fig. 11 a, b) and shift of the maximum in the EQE(I) dependence to low currents (Fig. 11 c, d). These effects are most pronounced for the certain

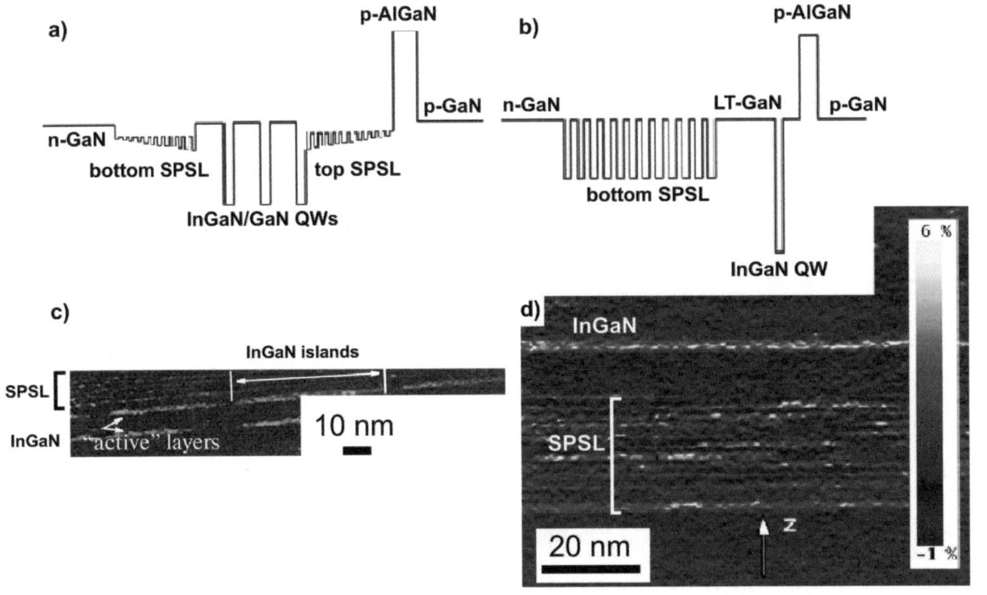

Fig. 10. Schemes (a, b) and HRTEM images (c, d) of the active region of blue (a, c) and deep green (b, d) LED structures based on combination of the InGaN/GaN SPSL and InGaN QWs (a). (Tsatsulnikov et al., 2010b; Lundin et al., 2011b)

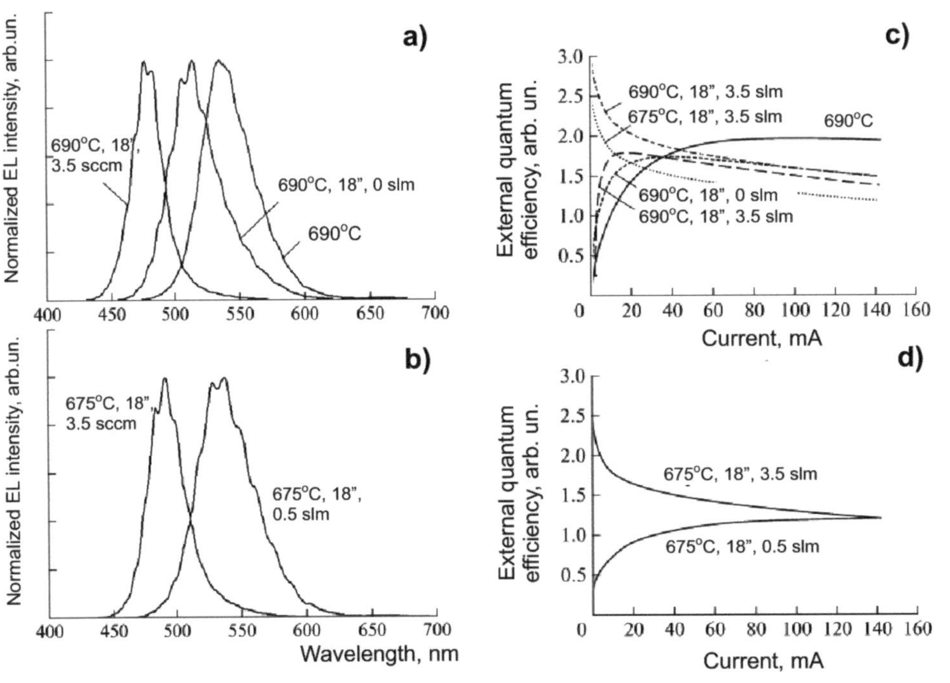

Fig. 11. EL spectra (a, b) and current dependence of the EQE (c, d) of the LED structures grown with GIs having different parameters (growth temperature, duration of the growth interruptions, hydrogen flow). (Tsatsulnikov et al., 2011a)

growth conditions. It is seen that on the one wafer either a slow monotonic increase in EQE with the current or its rapid increase in the region of low currents with a subsequent drop with the increase in the current are observed. This effect is associated with changes in the density of local In-rich QDs which were observed in the HRTEM images by following manner. Morphological transformation of the InGaN QW to array of QDs leads to variation in the energy spectrum of electrons and holes. An increase in the structural quality of the InGaN QW grown with the use of GIs, a decrease in the density of QDs, and improvement of the carrier transport in such transformed InGaN QWs lead to rapid population of the states of remaining QDs, which causes an abrupt short wavelength shift of the emission line with an increase in the current and a considerable increase in the EQE in the range of low currents (I ~5-20 mA for different samples). With further increase in the current, the states of the QDs become completely occupied and, due to effective transport, the fraction of the carriers reach region of dislocations propagating from the buffer layer that leads to a drop in the EQE with an increase in the current. A weak current dependence of the position of the emission maximum at the currents more than ~20 mA (Tsatsulnikov et al., 2011a) indicates that density of the QDs is not so significant to shift emission wavelength.

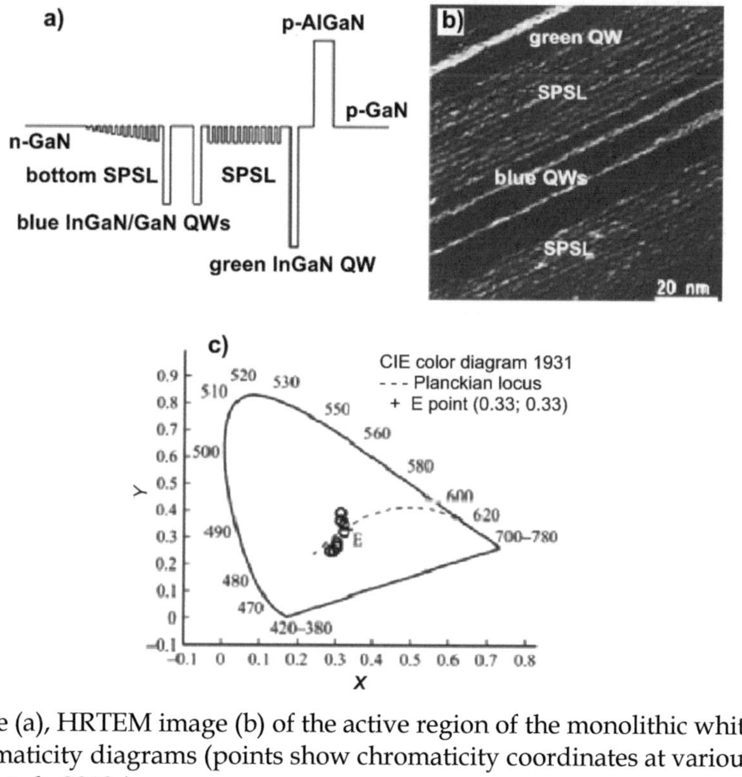

Fig. 12. Scheme (a), HRTEM image (b) of the active region of the monolithic white LED. . c- 1931 CIE chromaticity diagrams (points show chromaticity coordinates at various currents). (Tsatsulnikov et al., 2010c)

The combination of the InGaN/GaN SPSL with InGaN QWs was used for growth of monolithic white LEDs (Tsatsulnikov et al., 2010c) containing in blue and deep green QWs the active region. The SPSL was used as barrier between the blue and deep green QWs due to it's good vertical carrier transport. Scheme and HRTEM image of the active region of the monolithic white LED is shown in the Fig. 12 a, b. Mixing of the emission with wavelengths of 440 nm and 560 nm leads to white light emission with color parameters depending on current (Fig. 12 c, circles corresponds to different currents - increase in the current leads to the redistribution of the intensities of the blue and green lines and changes color coordinates). Comparison of the properties of the monolithic white LEDs with GaN and InGaN/GaN SPSL barriers has shown that using of the SPSL allows realization of effective emission from the all active InGaN layers even in the case of significant (24 nm) distance between blue and green QWs. However, for narrow (~6 nm) barrier LEDs with GaN barrier show better efficiency. Maximal EQE of ~6-7% and correlated color temperature (CCT) in the range of 6000 to 10000 K were demonstrated from these LEDs.

4. Conclusion

We have given an overview of the methods of in-situ fabrication of InGaN QDs It was shown that important growth parameters like interaction of InGaN with hydrogen, growth temperature and pressure play key roles in the formation of the InGaN QDs and its properties. Applying of the GIs with hydrogen admixing after InGaN QW growth affects redistribution of In atoms inside QW with modification of sizes of the small nanometer-size In-rich QDs and formation of large isolated islands. Applying of the cycle GIs during growth of a thick InGaN layer allows formation of nanocomposite InGaN/GaN short-period superlattices. Increase in the growth pressure at deposition of InGaN QW leads to formation of the separate QDs, modification of emission spectra and changes in the EQE of the LED. InGaN QDs can be formed by growth of special layer sequence in the heterostructures. It was shown that InGaN separate islands are formed in the composite InAlN/GaN/InGaN heterostructures due to increasing of the phase separation effect in the InGaN QW stimulated by InAlN wide bandgap islands. Described methods may be used for fabrication of active regions of light-emitting devices and for growth of other types of epitaxial structures for various applications. These effects must be taken into account during MOVPE of III-Ns, as it can have a strong positive or negative effect depending on the desired structure.

5. Acknowledgment

The authors would like thank his colleagues Dr. A. Sakharov, Dr. E. Zavarin, Dr. A. Nikolaev, Dr. V. Sizov, Dr. A. Zakgeim for the fruitful joint investigations. The authors are grateful to Dr. N. Cherkashin for TEM investigations and useful discussions. Authors are indebted to Svetlana-Optoelectronic company for fabrication of LEDs and regional JRC 'Material science and characterization in advanced technologies' for structural characterization.

6. References

Bai, J.; Wang, Q.; Wang, T.; Cullis, A. & Parbrook, P. (2009.) Optical and microstructural study of a single layer of InGaN quantum dots. *Journal of Applied Physics*, V.105, No.5, (March 2009), pp. 053505-053509, ISSN 0021-8979

Bedair, S.; McIntosh, F.; Roberts, J.; Piner, E., Boutros, K. & El-Masry, N. (1997). Growth and characterization of In-based nitride compounds. *Journal of Crystal Growth*, V.178, No.1-2, (June 1997), pp. 32-44, ISSN 0022-0248

Cho, H.; Lee, J.; Sharma, N.; Humphreys, C. & Yang. G. (2001), Effect of growth interruptions on the light emission and indium clustering of InGaNOGaN multiple quantum wells. *Appl. Phys. Lett.,* Vol. 79, pp. 2594 – 2596, ISSN 0003-6951

Choi, S.-K.; Jang, J-M.; Yi, S.-H; Kim, J.-A & Jung, W.-G. (2007). Fabrication and characterization of self-assembled InGaN quantum dots by periodic interrupted growth. *Proceedings of SPIE.* V.6479, (February 2007), pp.64791F

Daele, B., Tendeloo, G.; Jacobs, K.; Moerman, I. & Leys, M. (2004). Formation of metallic In in InGaN/GaN multiquantum wells. *Appl. Phys. Lett.* Vol. 85, pp.4379-4381, ISSN 0003-6951

Gerthsen, D.; Hahn, E.; Neubauer, B.; Rosenauer, A.; Scho¨n, O.; Heuken, M. & Rizzi, A. (2000). Composition Fluctuations in InGaN Analyzed by Transmission Electron Microscopy. *Phys. Status Solid* Vol. 177, pp.145-155, ISSN: 1862-6300

Hytch, M.; Snoeck, E. & Kilaas, R. (1998). Quantitative measurement of displacement and strain fields from HREM micrographs. *Ultramicroscopy,* V.74, No.3, (August 1998), pp.131-146, ISSN 0304-3991

Ji, L.; Su, Y.; Chang, S.; Tsai, S.; Hung, S.; Chuang, R.; Fang, T. & Tsai, T. (2004). Growth of InGaN self-assembled quantum dots and their application to photodiodes. *Journal of Vacuum Science & Technology.* V.A22, (May 2004), pp.792-795, ISSN 0734-2101

Karpov, S.; Podolskaya, N.; Zhmakin, I. & Zhmakin, A. (2004). Statistical model of ternary group-III nitrides. *Physical Review B,* V.70, No.23, (December 2004), p. 235203 ISSN 1098-0121

Krestnikov, I.; Ledentsov, N.; Hoffmann, A.; Bimberg, D.; Sakharov, A.; Lundin, W.; Tsatsul'nikov, A.; Usikov,A.; Alferov, Zh.; Musikhin, Yu. & Gerthsen, D. (2002). Quantum dot origin of luminescence in InGaN-GaN structures. *Physical Review B,* V.192, (October 2002), pp. 155310, ISSN 1098-0121

Krestnikov, I.; Sakharov, A.; Lundin, W.; Usikov, A.; Tsatsulnikov, A.; Musikhin, Yu.; Gerthsen, D.; Ledentsov, N.; Hoffmann & A.; Bimberg, D. (2002). Time-Resolved Studies of InGaN/GaN Quantum Dots. *Physica Status Solidi,* V.192, No.1, pp. 49-53, ISSN 1862-6300

Kryzhanovskaya, N.; Lundin, W.; Nikolaev, A.; Tsatsul'nikov, A.; Sakharov, A.; Pavlov, M.; Cherkachin, N.; Hÿtch, M.; Valkovsky, G.; Yagovkina, M. & Usov, S. (2010). Optical and structural properties of InGaN/GaN short-period superlattices for the active region of light- emitting diodes. *Semiconductors,* V.44, No.6, pp.828-834, ISSN 1063-7826

Kumar, M.; Park, J.; Lee, Y.; Chung, S.; Hong, Ch. & Suh, E.. (2008). Improved Internal Quantum Efficiency of Green Emitting InGaN/GaN Multiple Quantum Wells by In Preflow for InGaN Well Growth. *Japanese Journal of Applied Physics* V.47, No.2, (2008), pp. 839–842, ISSN 0021-4922

Liu, W.; Chua, S.; Zhang, X. & Zhang, J. (2003). Effect of high temperature and interface treatments on photoluminescence from InGaN/GaN multiple quantum wells with green light emissions. *Applied Physics Letters,* V.83, No.5, (August 2003), pp.914-916, ISSN 0003-6951

Lundin, W.; Zavarin, E. & Sizov, D. (2005). *Technical Physics Letters.* Influence of the Carrier Gas Composition on Metalorganic Vapor Phase Epitaxy of Gallium Nitride. Vol. 31, No. 4, 2005, pp. 293-294. ISSN: 1063-7850

Lundin, W.; Sakharov, A.; Zavarin, E.; Sinitsyn, M.; Nikolaev, A.; Mikhailovsky, G.; Brunkov, P.; Goncharov, V.; Ber, B.; Kazantsev, D. & Tsatsulnikov, A. (2009) Effect of Carrier Gas and Doping Profile on the Surface Morphology of MOVPE Grown Heavily Doped GaN:Mg Layers *Semiconductors,* Vol 43 No. 7 (July 2009) pp. 963 - 967. ISSN: 1063-7826

Lundin, W.; Zavarin, E.; Sinitsyn, M.; Nikolaev, A.; Sakharov, A.; Tsatsulnikov, A.; Yakovlev, E.; Talalaev, R.; Lobanova, A. & Segal, A. (2009), Optimization of III-N heterostructures growth by MOVPE via surface processes control. *Proceedings of the 13th European Workshop on Metalorganic Vapour Phase Epitaxy,* pp. 9-14 Ulm, Germany, 7-10 June 2009.

Lundin, W.; Zavarin, E.; Sinitsyn, M.; Sakharov, A.; Usov, S.; Nikolaev, A.; Davydov, D.; Cherkashin, N. & Tsatsulnikov, A. (2010). Effect of pressure in the growth reactor on the properties of the active region in the InGaN/GaN light-emitting diodes. *Semiconductors,* V.44, No.1, pp.123-126, ISSN 1063-7826

Lundin, W.; Sakharov, A.; Tsatsulnikov, A. & Ustinov, V. (2011) MOVPE of device-oriented wide-band-gap III-N heterostructures. *Semicond. Sci. Technol.* Vol. 26 No. 1 (January 2011) p. 014039 ISSN: 0268-1242

Lundin, W.; Nikolaev, A.; Sakharov, A.; Zavarin, E.; Valkovskiy, G.; Yagovkina, M; Usov, S.; Kryzhanovskaya, N.; Sizov, V.; Brunkov, P.; Zakgeim, A.; Cherniakov, A.; Cherkashin, N.; Hytch, M.; Yakovlev, E.; Bazarevskiy, D.; Rozhavskaya, M. & Tsatsulnikov, A. (2011). Single quantum well deep-green LEDs with buried InGaN/GaN short-period superlattice. *Journal of Crystal Growth,* V.315, pp.267–271, ISSN 0022-0248

Martin, R.; Middleton, P.; O'Donnell, K. & Stricht, W. (1999). Exciton localization and the Stokes' shift in InGaN epilayers. *Applied Physics Letters,* V.74, No.2, (January 1999), pp. 263-265, ISSN 0003-6951

Moon, Y.-T.; Kim, D.-J.; Song, K.-M.; Choi, Ch.-J.; Han, S.-H.; Seong, T.-Y. & Park, S.-J. (2001). Effects of thermal and hydrogen treatment on indium segregation in InGaN/GaN multiple quantum wells. *Journal of Applied Physics,* V.89, No.11, (June 2001), pp. 6514-6518, ISSN 0021-8979

Mukai, T.; Yamada, M. & Nakamura, Sh. (1999) Characteristics of InGaN-Based UV/Blue/Green/Amber/Red Light-Emitting Diodes. *Jpn. J. Appl. Phys.* Vol. 38 Part 1, No. 7A (July 1999) pp. 3976–3981 ISSN: 0021-4922

Musikhin, Yu.; Gerthsen, D.; Bedarev, D.; Bert, N.; Lundin, W.; Tsatsul'nikov, A.; Sakharov, A.; Usikov, A. Alferov, Zh.; Krestnikov, I.; Ledentsov, N.; Hoffmann, A. & Bimberg, D. (2002). Influence of metalorganic chemical vapor deposition growth conditions on In-rich nanoislands formation in InGaN/GaN structures. *Applied Physics Letters,* Vol.80, No.12, (March 2002), pp. 2099-2101, ISSN 0003-6951

Nakamura, S.; Senoh, M. & Mukai, T. (1993) P-GaN/N-InGaN/N-GaN Double-Heterostructure Blu-Light-Emitting_Diodes. *Jpn. J. Appl. Phys.* V 32 No 1 (January 1993), pp. L8-L11. ISSN: 0021-4922

Oliver, R.; Briggs, G.; Kappers, M; Humphreys, C.; Yasin, Sh.; Rice, J.; Smith, J. & Taylor R. (2003). InGaN quantum dots grown by metalorganic vapor phase epitaxy employing a post-growth nitrogen anneal. *Applied Physics Letters,* Vol.83, No. 4, (July 2003), pp. 755-757, ISSN 0003-6951

Paillard, M.; Marie, X.; Vanelle, E.; Amand, T. ; Kalevich, V.; Kovsh, A.; Zhukov, A. & Ustinov, V. (2000). Time-resolved photoluminescence in self-assembled InAs/GaAs quantum dots under strictly resonant excitation. *Applied Physics Letters,* Vol.76, No.1, (January 2002), pp. 76-78, ISSN 0003-6951

Piner, E.; Behbehani, M.; El-Masry, N.; McIntosh, F.; Roberts, J.; Boutros, K. & Bedair, S. (1997), Effect of hydrogen on the indium incorporation in InGaN epitaxial films. *Appl. Phys. Lett.* Vol. 70 No. 4 (January 1997), pp. 461-463. ISSN 1063-7826

Rosenauer, A.; Kaiser, S.; Reisinger, T.; Zweik, J.; Gebhard, W. & Gerthsen, D. (1996), Digital analysis of high-resolution transmission electron microscopy lattice images, *Optik,* Vol. 102, pp.63-69. ISSN: 0030-4026

Sakharov, A.; Lundin, W.; Krestnikov, I.; Bedarev, D.; Tsatsul'nikov, A.; Usikov, A.; Alferov, Zh.; Ledentsov, N.; Hoffmann, A. & Bimberg, D. (2000). Influence of Growth Interruptions and Gas Ambient on Optical and Structural Properties of InGaN/GaN Multilayer Structures. *Proceedings of IWN 2000 4th International Workshop on Nitride semiconductors,* pp. 241-243, ISBN 842-6508-23-3, Nagoya, Japan, September 24-27, 2000

Sakharov, A.; Lundin, W.; Zavarin, E.; Sinitsyn, M.; Nikolaev, A.; Usov, S.; Sizov, V.; Mikhailovsky, G.; Cherkashin, N.; Hytch, M.; Hue, F.; Yakovlev, E.; Lobanova, A. & Tsatsulnikov, A.(2009) Effect of strain relaxation on active-region formation in InGaN/(Al)GaN heterostructures for green LEDs. *Semiconductors,* v.43, No.6 pp.812-817, ISSN 1063-7826

Schenk, H.; Mierry, P.; Lau¨gt, M.; Omne`s, F.; Leroux, M.; Beaumont, B. & Gibart, P. (1999). Indium incorporation above 800 °C during metalorganic vapor phase epitaxy of InGaN. *Applied Physics Letters,* V.75, No.17, (October 1999), pp. 2587-2589, ISSN 0003-6951

Shim, H.; Choi, R.; Jeong, S.; Vinh, L.; Hong, C.-H.; Suh, E.-K.; Lee, H.; Kim, Y.-W. & Hwang, Y. (2002). Influence of the quantum-well shape on the light emission characteristics of InGaN/GaN quantum-well structures and light-emitting diodes. *Applied Physics Letters,* V.81, No.19, (November 2002), pp.3552-3554, ISSN 0003-6951

Soh, C.; Liu, W.; Teng, J.; Chow, S.; Ang, S. & Chua. S. (2008). Cool white III-nitride light emitting diodes based on phosphor-free indium-rich InGaN nanostructures. *Applied Physics Letters,* V.92, No.26, (January 2008) pp.261909-261911, ISSN 0003-6951

Sun, Y.; Choa, Y.-H.; Suh, E.-K.; Lee, H.; Choi, R. & Hahn, Y. (2004). Carrier dynamics of high-efficiency green light emission in graded-indium-content InGaN/GaN quantum wells: An important role of effective carrier transfer. *Applied Physics Letters,* V.84, No.1, (January 2004), pp.49-51, ISSN 0003-6951

Takeuchi, M.; Maegawa, T.; Shimizu, H.; Ooishi, S.; Ohtsuka, T. & Aoyagi, Y. (2009), AlN/AlGaN short-period superlattice sacrificial layers in laser lift-off for vertical-type AlGaN-based deep ultraviolet light emitting diodes. *Appl. Phys. Lett.* Vol. 94, pp. 61117-1 – 61117-3

Tsatsul'nikov, A.; Zavarin, E.; Kryzhanovskaya, N.; Lundin, W.; Saharov, A.; Usov, S.; Brunkov, P.; Goncharov, V.; Cherkashin, N. & Hytch, M. (2010). Formation of composite InGaN/GaN/InAlN quantum dots. *Semiconductors,* V.44, No.10, pp.1338-1341, ISSN 1063-7826

Tsatsulnikov, A.; Lundin, W.; Sakharov, A.; Zavarin, E.; Usov, S.; Nikolaev, A.; Cherkashin, N.; Ber, B.; Kazantsev, D.; Mizerov, M.; Park, H.; Hytch, M. & Hue, F. (2010). Active region based on graded-gap InGaN/GaN superlattices for high-power 440- to 470-nm light-emitting diodes. *Semiconductors,* V.44, No.1, pp.93-97, ISSN 1063-7

Tsatsulnikov, A.; Lundin, W.; Sakharov, A.; Zavarin, E.; Usov, S.; Nikolaev, A.; Kryzhanovskaya, N.; Synitsin, M.; Sizov, V.; Zakgeim & A.; Mizerov, M. (2010). A monolithic white LED with an active region based on InGaN QWs separated by short-period InGaN/GaN superlattices. *Semiconductors,* V.44, No.6, pp.808-811, ISSN 1063-7826

Tsatsulnikov, A.; Lundin, W.; Zavarin, E.; Nikolaev, A.; Sakharov, A.; Sizov, V.; Usov, S.; Musikhin, Yu. & Gerthsen, D. (2011). Influence of hydrogen on local phase separation in InGaN thin layers and properties of light-emitting structures based on them. *Semiconductors,* V.45, No.2, pp.271-276, ISSN 1063-7826

Tsatsulnikov, A.; Lundin, W.; Sakharov, A.; Zavarin, E.; Usov, S.; Nikolaev, A.; Kryzhanovskaya, N.; Sizov, V.; Synitsin, M.; Yakovlev, E.; Chernyakov, A.; Zakgeim, A.; Cherkashin, N. & Hytch, M. (2011). InGaN/GaN short-period superlattices: synthesis, properties, applications. *Physica Status Solidi C,* V.8, No.7-8, pp. 2308-2310, ISSN 1862-6300

Wang, Q.; Wang, T.; Bai, J.; Cullis, A.; Parbrook P. &Ranalli, F. (2008). Growth and optical investigation of self-assembled InGaN quantum dots on a GaN surface using a high temperature AlN buffer. *Journal of Applied Physics,* V.103, No.12, (June 2008), pp.123522-123528, ISSN 0021-8979

Wen, T.-Ch.; Lee, Sh-Ch. & Lee W-I. (2001). Light-Emitting Diodes: Research, Manufacturing, and Applications. *Proceedings of SPIE,* V.4278, (May 2001), pp. 141-149

Yakovlev, E.; Talalaev, R.; Kondratyev, A.; Segal, A.; Lobanova, A.; Lundin, W.; Zavarin, E.; Sinitsyn, M.; Tsatsulnikov, A. & Nikolaev, A. (2008). Hydrogen effects in III-nitride MOVPE. *Journal of Crystal Growth.* V.310, pp. 4862-4866. ISSN 0022-0248

Yong-Tae Moon, Dong-Joon Kim, Keun-Man Song, Chel-Jong Choi, Sang-Heon Han, Tae-Yeon Seong, Seong-Ju Park. (2001). Effects of thermal and hydrogen treatment on indium segregation in InGaN/GaN multiple quantum wells. *Journal of Applied Physics.* Vol. 89, No.11, (June 2001), pp. 6514-6518, ISSN 0021-8979

Yoshimoto, N.; Matsuoka, T.; Sasaki, T. & Katsui, A. (1991). Photoluminescence of InGaN films grown at high temperature by metalorganic vapor phase epitaxy. *Applied Physics Letters,* V.59, No.18, (October 1991), pp. 2251-2253, ISSN 0003-6951

Zavarin, E.; Sizov, D.; Lundin, W.; Tsatsulnikov, A.; Talalaev, R.; Kondratyev, A. & Bord, O. (2005), *In-Situ* investigations of GaN chemical unstability during MOCVD. *Proceedings of the EUROCVD-15,* vol. 2005-09, Bochum, Germany, 2005, Electrochemical Society, NJ, USA, pp. 299-305

10

Room Temperature Synthesis of ZnO Quantum Dots by Polyol Methods

Rongliang He and Takuya Tsuzuki
Centre for Frontier Materials
Deakin University, Geelong Technology Precinct, Geelong, VIC, Australia

1. Introduction

Zinc oxide is an important semiconducting material having a broad range of applications including transparent conductive oxides (Hosono 2007), ultraviolet (UV) light absorbers (Becheri et al. 2008) and photocatalysts (Beydoun et al. 1999). ZnO nanoparticles exhibit quantum size effects when the particle size is smaller than the exciton-Bohr diameter of ~ 8 nm. In the past, many methods to synthesize ZnO quantum dots were investigated, (Zhang et al. 2010; Xiong et al. 2005; Tang et al. 2009; Tsuzuki & McCormick, 2001). Among them, wet chemical synthesis methods, such as sol-gel, hydrothermal and solvothermal methods, are widely used to obtain free ZnO quantum dots that are detached from substrates. However, this approach normally requires surfactants or capping agents to limit the growth of particles below the exciton-Bohr diameter. This causes some drawbacks such as the necessity to have additional steps in the production process to remove the surfactants from nanoparticles and increased chance of contamination. In addition, wet chemical synthesis of ZnO normally requires high reaction temperatures above 100 °C. For example, Tang et al. showed that mono-dispersed ZnO quantum dots can be synthesised in triethylene glycol and demonstrated their promising applications in cell labelling, but the reaction required a temperature as high as 280 °C and stearate acid as a surfactant (Tang et al. 2009). When the high temperatures were not used, the reaction takes a considerable time. Xiong et al. synthesised ZnO quantum dots in triethylene glycol using LiOH and zinc acetate and, although the morphology and optical properties of ZnO was controlled by the molar ratio between LiOH and zinc acetate, this reaction required 30 days to complete at the interface between the air and the polyol solution (Xiong et al. 2011).

This chapter reports a series of novel room-temperature methods to synthesize mono-dispersed ZnO quantum dots in a polyol without using any surfactant additives. Tetraethylene glycol (TEG) was used as the solvent system for the following reasons. TEG is a clear, non-hazardous, inexpensive organic liquid which is miscible with water. TEG has a structural formula of $HO\text{-}CH_2\text{-}CH_2(\text{-}O\text{-}CH_2\text{-}CH_2)_3\text{-}OH$ and is sufficiently polar to allow many inexpensive salt raw materials to be dissolved. Hence TEG is more economical than other organic solvents to be used for the synthesis. Furthermore, TEG is reported to have much lower affinity to ZnO surface than PEG or other polar solvents, which is advantageous for the separation of resulting nanoparticles from the solvent phase by simple washing with deionised water (Dobryszycki & Biallozor 2001).

In this research, NaOH and $ZnCl_2$ were used as raw materials and were directly added into TEG without any surfactant additives. In an aqueous environment, $ZnCl_2$ and NaOH normally react readily to form $Zn(OH)_2$ precipitates. However, it was found that the mixture of $ZnCl_2$ and NaOH solutions in TEG did not induce the reaction at room temperature without external energy input. In our previous research (He & Tsuzuki 2010), it was found that, the dissolution speed of the $ZnCl_2$ and NaOH precursors in TEG was quite low at room temperature. Higher temperatures around 60~70 °C were necessary to increase the dissolution speed of the precursors. In this study, it was demonstrated that mechanical agitation and UV light irradiation can be used to initiate the reaction to form ZnO quantum dots at room temperature. It was also found that the size of the quantum dots can be tailored by controlling the process parameters, without additional surfactants or capping agents to limit the particle growth.

The details about the effect of process parameters on particle size, physical properties (bandgap, photoluminescence, etc.), and the growth mechanism of ZnO quantum-dots are presented in the following sections.

2. Mechanically induced room temperature synthesis of ZnO quantum dots in TEG

This section describes the activation of the formation of ZnO quantum dots by mechanical milling at room temperature.

2.1 Experimental

In a typical synthesis, 30 ml of tetraethylene glycol (TEG, 99%, Sigma-Aldrich Ltd) was placed in a sealed plastic container with 0.952g $ZnCl_2$ (98%, Fluka, Sigma-Aldrich Ltd., NSW, Australia), 0.56g NaOH pellets (97%, Chem-Supply, SA, Australia) and 10g ZrO_2 milling balls (0.8-1 mm). Then, the precursors were milled for 2 hours using a Spex 8000 mixer/mill to form ZnO quantum dots at room temperature. The prepared quantum dots were sedimented and washed with ethanol using a centrifuge at the speed of 7000 rpm (Eppendorf centrifuge 5417R) until the salinity of the supernatant becomes less than 100 ppm.

The morphology of the ZnO-TEG colloid was studied by transmission electron microscopy (TEM), using a Jeol 2100 microscope with the beam energy of 200 kV. TEM specimen was prepared by evaporating a drop of nanoparticle dispersion on a carbon-coated specimen grid. Particle size distribution was measured by a dynamic light scattering (DLS) method using a Malvern Zetasizer-Nano instrument. UV-vis spectra were obtained using a Varian Cary 3E spectrophotometer. Photoluminescence of ZnO quantum dots was measured using a Varian Cary Eclipse fluorescence spectrometer. For TEM, UV-vis spectroscopy, photoluminescence spectroscopy and DLS measurements, as-prepared ZnO-TEG colloid was used in an diluted form without washing to remove reaction by-products.

The crystal phase and crystallite size were studied by powder X-ray diffraction (XRD) on an X-ray diffractometer (Panalytical X'Pert PRO MRD) with Cu-Kα radiation at a step width of 0.02° (λ=1.54180 Å). The presence of TEG on the ZnO quantum dots was investigated by Fourier transform infrared spectroscopy (FTIR) using a Bruker Vertex 70 instrument. For XRD and FTIR measurements, washed and dried quantum dots were used.

2.2 Results and discussions

The raw materials, NaOH and $ZnCl_2$ powders, were not readily soluble in TEG at room temperature without mechanical energy input. After the ball milling process, NaOH and $ZnCl_2$ powders were completely dissolved and no sedimentation was observed in the TEG solution. The pH value of the solution was around 8 after 2 hours of milling, which indicated the consumption of OH^- during milling through the reaction between Zn and OH ions. The solution was yellowish and visually transparent. However, when a laser beam was shone onto the as-milled sample, the Tyndall effect was displayed as shown in Figure 1(a), indicative of the presence of quantum dots in the TEG. The bright-field TEM image shown in Figure 1(b) revealed that the size of the synthesized ZnO quantum dots were around 3-5 nm. However, most of the quantum dots appeared to form clusters on the TEM grid. Further high-resolution TEM study shown in Figure 1(c) revealed that the lattice fringes of the particles run in the same direction across the whole structure, indicating that the single quantum dot possessed good crystallinity. Most of the quantum dots were less than 5 nm. The results of particle sizing using the DLS technique support the TEM observation (Figure 1(d)). In the TEG-diluted solution of the as-milled sample, over 90% of the quantum dots were under 5 nm in diameter and average particle size was 3.1 nm.

The crystal structure was also confirmed by XRD (Figure 2) where only the wurtzite phase corresponding to the standard crystallographic data in the JCPDS-ICDD index card No. 36-1451 was observed, indicating that the synthesized powder consisted of a single phase. The crystal size calculated from the diffraction peak broadening using the Scherrer equation was around 4.6 nm, in good agreement with the result of TEM and DLS studies.

The ZnO quantum dots showed significantly high dispersion stability in the TEG solution even after a long aging time; no precipitation was observed after 1 month of aging in the dark environment. In Figure 3(a), it is shown that both fresh and 1-month-aged sample showed identical yellowish colour and transparency. Furthermore, DSL measurements showed that the growth speed of ZnO quantum dots during aging is extremely slow. The particle size increased to only 8-10 nm after 1 month of aging.

It was found that the as-prepared ZnO TEG solution is stable in the glycols but not in water or ethanol. To investigate the dispersion stability of ZnO quantum dots, 30 μl of as-prepared solution was added into 10 ml TEG, ethylene glycol, water and ethanol separately. Then, UV-Vis absorbance measurements were carried out to these diluted ZnO suspensions. In Figure 4, the light scattering effect appeared in the water and ethanol solutions as the higher absorbance values in the visible light range, which indicates the agglomeration of the ZnO quantum dots. However, for the ethylene glycol and TEG solutions, no such absorption or scattering effect was observed in the visible light range, which implies that the quantum dots were well dispersed in the solution without agglomeration.

The absorbance peak position and intensity also varied in the different solutions which revealed the variation of bandgap energy. According to the Tauc's equation, the band gap energy of the nanoparticles can be estimated from the absorbance spectra for direct bandgap semiconductors using the following equation (Tauc et al. 1966):

$$(\alpha h\nu)^2 \propto h\nu - E_g \quad (1),$$

where α is the absorption coefficient, $h\nu$ is the photon energy and Eg is the Tauc optical bandgap energy. By extrapolating the linear portion of the plot $(\alpha h\nu)^2$ versus $h\nu$ to the

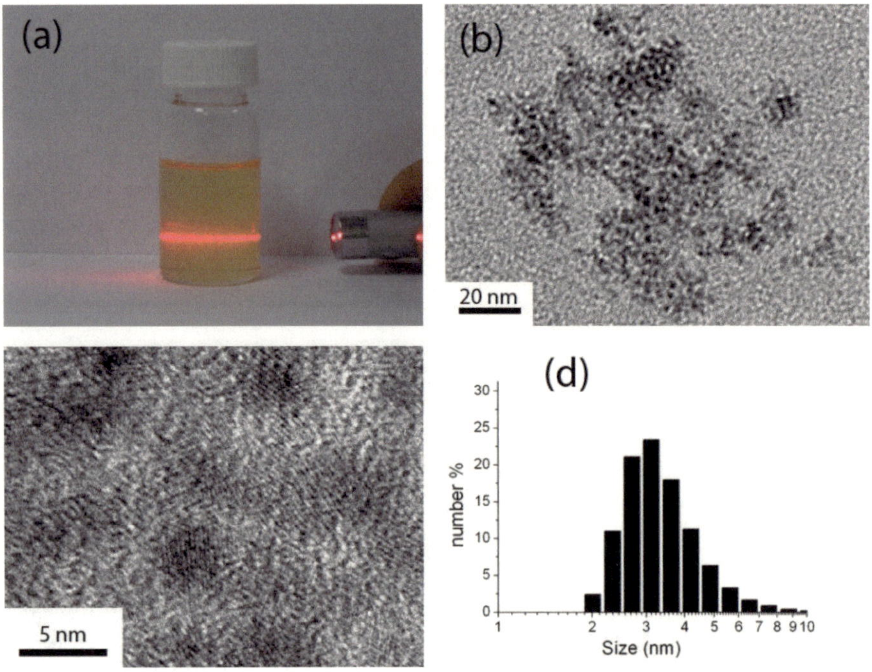

Fig. 1. (a) The Tyndall effect on as-prepared ZnO-TEG solution; (b) bright-field TEM image of as-prepared ZnO; (c) high-resolution TEM image of as-prepared ZnO; (d) size distribution of ZnO in TEG solution.

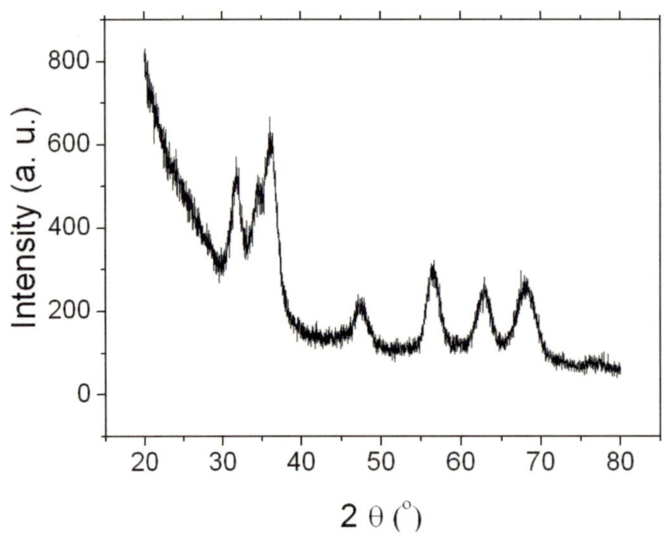

Fig. 2. XRD patterns of washed ZnO.

$(ahv)^2 = 0$ axis, the bandgap value was calculated. The insert graph shows that the band gap of ZnO quantum dots is 3.525 eV in water and 3.556 eV in ethanol. These values are smaller than that of ZnO in ethylene glycol and TEG (3.636 and 3.644 eV respectively), because of the weakening of the quantum effect due to agglomeration (Tong et al. 2011). The results also support the observation that the as-prepared ZnO quantum dots have high dispersion stability in the polyol medium.

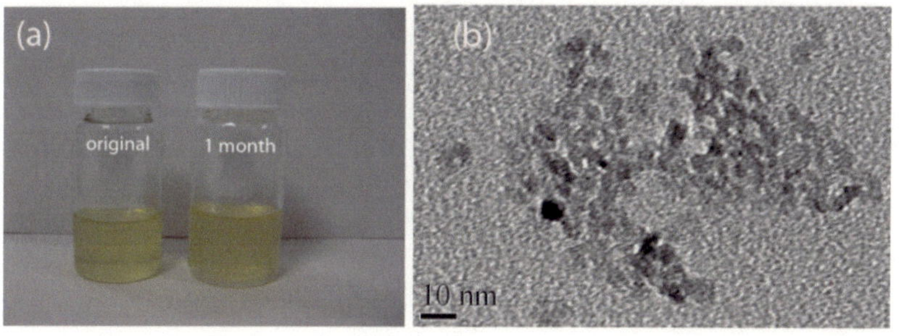

Fig. 3. (a) Appearance of original and 1 month aged ZnO-TEG colloid, and (b) bright-field TEM image of 1 month aged ZnO quantum dots.

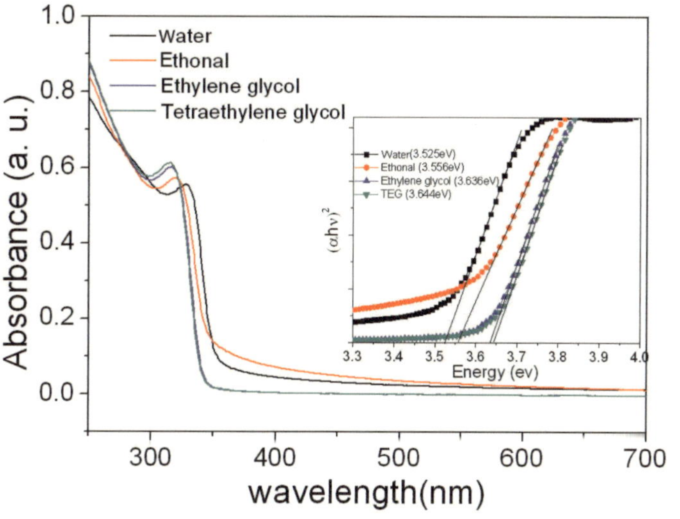

Fig. 4. UV-vis absorbance spectra of ZnO in different solutions. The inset depicts the results of band-gap energy calculation.

The washed sample was also characterized by FTIR to identify the functional groups on the particles to explain its high dispersion stability in TEG. Figure 5 shows FTRI spectra of TEG and washed ZnO quantum dots. The broad peaks at 3420 and 1620 cm^{-1} correspond to the O-H stretching and bending vibrations, respectively, resulting from the presence of hydroxyl groups on the particle surface. The peaks at 2924 and 1384 cm^{-1} are attributed to the unsymmetrical stretching and bending vibrations of $-CH_2$ groups in glycol, respectively. The other characteristic peaks of TEG at 1089.4 and 888.6 cm^{-1} were also present in the spectrum of ZnO nanoparticles (Vafaee & Ghamsari 2007). The results suggest that the ZnO quantum dots were capped by TEG molecules even after extensive washing with water to remove TEG. The TEG molecules on the surface of quantum dots may provide steric hindrance to provide high stability in polyol solutions such as ethylene glycol and TEG. When the colloid was diluted by water or ethanol, the Van der Waals force would assemble the quantum dots together to exhibit turbidity.

In our previous research (He & Tsuzuki 2010), it was found that, at room temperature, the $ZnCl_2$ and NaOH solutions in TEG did not cause reaction with each other at room temperature and that over 100 °C was required to form ZnO. In order to elucidate the cause of ZnO formation at room temperature by mechanical milling, different amounts of raw materials, that is, a higher concentration (sample A) and a lower concentration (sample B) than the sample described above, were used in this polyol-mechanochemical process as shown in Table 1.

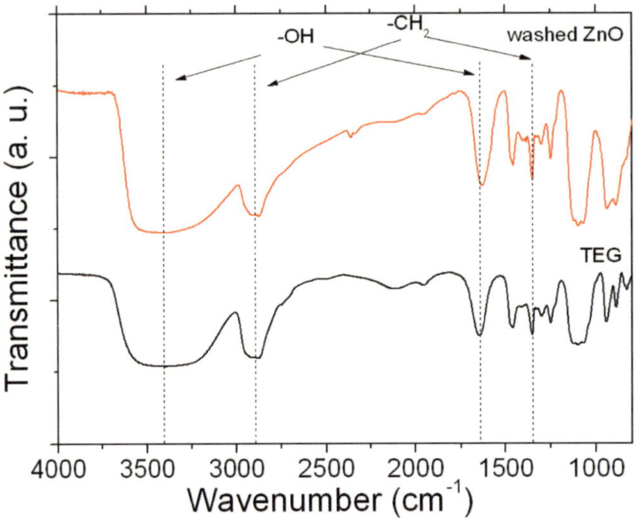

Fig. 5. FT-IR spectra of TEG and washed ZnO.

Sample ID	$ZnCl_2$(g)	NaOH(g)	TEG(ml)
A	1.36	0.8	30
B	0.408	0.24	30

Table 1. Components of raw materials.

After the ball milling process, both solutions of sample A and sample B exhibited a yellowish and transparent appearance. Then, UV-vis absorbance spectra and photo-luminescence spectra were taken on the sample A and sample B (30μl - 10ml) without washing to remove the reaction by-products. Figure 6 shows the UV-vis absorbance spectra of diluted sample A and sample B. It is evident that only sample A has the typical ZnO absorption peak around 340 nm. No absorption peaks can be observed for sample B, indicating that no ZnO quantum dot was synthesised after 2 hours of mechanical

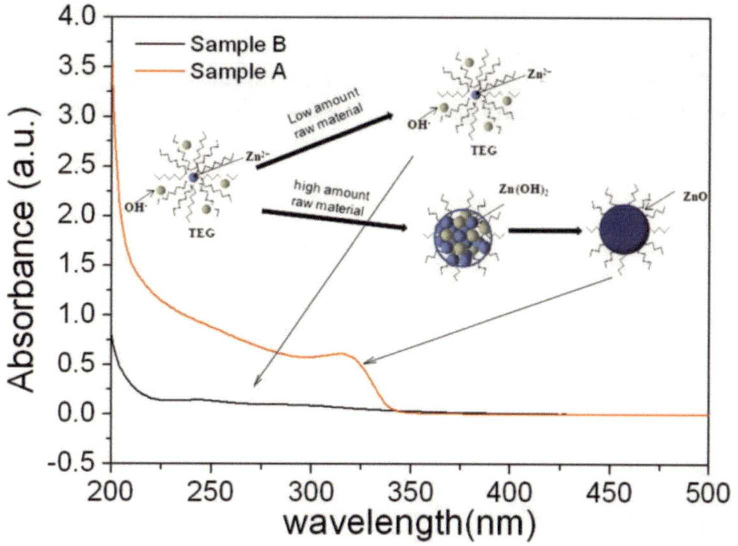

Fig. 6. UV-vis absorbance spectra of diluted sample A and sample B. The inset depicts the reaction process.

milling. Hence, only high concentration of raw materials in the solution resulted in the synthesis of ZnO quantum dots at room temperature using mechanical energy input.

The schematic illustration inserted in Figure 6 gives the possible reaction mechanism involved in the synthesis process. Generally, the hydroxyl groups at both ends of a glycol molecule act as electron donors, giving their lone-pair electrons to those that are electron deficient such as a metal cation, leading to the formation of stable Zn-TEG chelates (Larcher et al. 2003). The formation of stable zinc chelates prevents Zn ions from directly react with OH^- to form $Zn(OH)_2$ at room temperature without other external energy input.

When mechanical energy was introduced to the system, Zn-TEG chelates were de-stabilised and Zn ions are released from the chelates to react with OH^- to form ZnO. When the quantity of raw materials in the TEG solution is lower, the released Zn ions tend to form Zn-TEG chelate again before reacting with OH^-, due to the lesser chance to encounter OH^- in the very close proximity of original Zn-TEG chelates and the strong affinity of TEG to form chelates. This model, however, does not explain why the pH of the as-milled solution of sample B is nearly neutral despite the lack of the formation of ZnO, which is a subject of future study.

In Figure 7, photoluminescence emission spectra of sample A and sample B is shown. Sample A had an emission peak at 525 nm while the emission peak of sample B appeared at 472 nm. The inset photo in Figure 7 shows the difference in the colour of the emitted light, yellow for sample A and blue for sample B. The yellow emission at 500–550 nm was also observed by other research groups who prepared ZnO quantum dots by sol-gel methods (Xiong 2010; Wang et al. 2006; Goharshadi et al. 2011). According to the van Dijiken model, the yellow emission light in sample-A stemmed from the transition of an electron from the conduction band to a deep bandgap level associated with surface defects or oxygen vacancies (van Dijken et al. 2000). ZnO was not present in sample B. Hence the blue emission the blue emission may have arisen from the Zn-TEG chelates and the photoluminescence may be related to metal assisted ligand-to-ligand charge transfer and an intra-ligand π-π* transition (Gronlund et al. 1995; Bauer et al. 2007), which should be confirmed by further study.

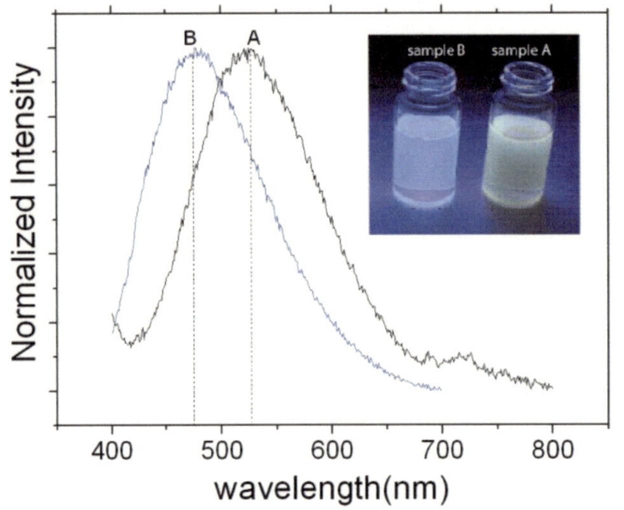

Fig. 7. Photo luminescence emission spectra of diluted sample A and sample B. The inset depicts the colours or emission light under UV light irradiation.

3. UV induced room temperature synthesis of ZnO quantum dots in TEG

This section describes the activation of the formation of ZnO quantum dots by UV irradiation at room temperature.

3.1 Experimental

The synthesis of Zn-TEG chelates was performed firstly by mixing 30 ml of TEG (99%, Sigma-Aldrich Ltd), 0.408g $ZnCl_2$ (98%, Fluka, Sigma-Aldrich Ltd., NSW, Australia), 0.24g NaOH pellets (97%, Chem-Supply, SA, Australia) and 10g ZrO_2 milling balls (0.8-1 mm) together and then ball milling the mixture in a Spex 8000 mixer for 2 hours. This is the same procedure as for preparing the sample B in the previous section. Then, UV-light was irradiated on the as-prepared Zn-TEG chelates solution with continuous stirring in an ice bath. The as-prepared ZnO-TEG colloidal suspension was directly diluted for the characterization by transmission electron microscopy (TEM), UV-vis spectroscopy, photo-luminescence spectroscopy and dynamic light scattering (DLS) measurements.

3.2 Results and discussions

Figure 8 shows the UV-vis absorbance spectra of the solution that was irradiated for up to 8 hours. It is evident that the absorption peak around 300-350 nm gradually appeared when the exposure time was prolonged over 2 hours, indicative of the formation of ZnO quantum dots. For the sample that was irradiated for 2 hours, the absorption peak appeared at 300 nm which is a much smaller wavelength than that of bulk ZnO. After irradiation for 4 hours, the peak intensity nearly doubled and the peak position was red-shifted to 316 nm, which is attributed to the increase of the number of nucleated ZnO quantum dots and the further growth in particle size. After 8 hours of UV exposure, the peak intensity did not increase much, while the peak position was further red-shifted to 324 nm. The results indicate that the formation of ZnO quantum dots had completed after 4 hours of UV irradiation and that agglomeration or particle growth occurred afterwards. The inserted photo in Figure 8 was taken under UV-light to show the colour of photoluminescence emission. The high intensity of green-yellow emission under UV-light irradiation was an evidence for the presence of ZnO quantum dots and their excellent photoluminescence properties.

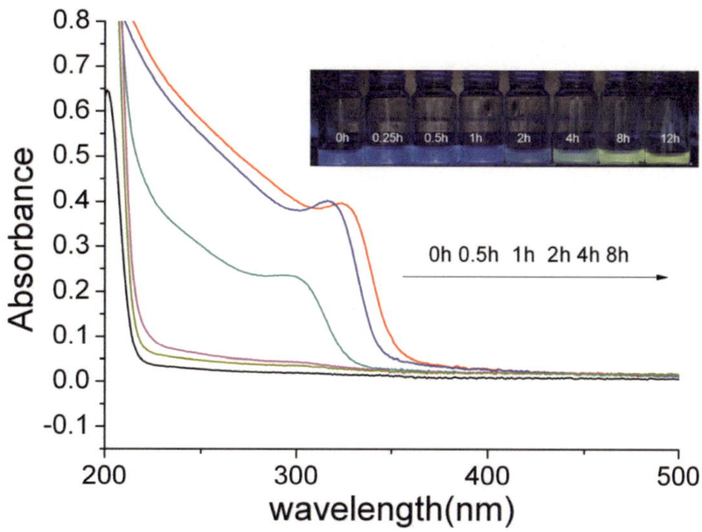

Fig. 8. UV-Vis absorbance spectra of UV-light irradiated TEG solution. The inset depicts the colours of emission light under UV light irradiation.

Because of the strong quantum size effect on the absorption peak energy, the average particle diameter of the ZnO quantum dots can be estimated using the following equation based on the effective mass model (Kumbhakar et al. 2008):

$$d = 2 \times \frac{-0.3049 + \sqrt{-26.23012 + \frac{10240.72}{\lambda_p}}}{-6.3829 + \frac{2483.2}{\lambda_p}} \quad (2),$$

where λ_p is the optical bandgap wavelength. The average diameter thus calculated was 2.6 nm, 3.0 nm and 3.2 nm for the samples irradiated for 2 hours, 4 hours and 8 hours, respectively.

The estimated particle sizes are in good agreement with the sizes observed under TEM (Figure 9). For the sample irradiated for 2 hours, mono-dispersed ZnO nanoparticles were found to have sizes of ~2 nm. These particles had a low contrast in the TEM images against the carbon film background, possibly due to their low crystallinity. For the sample irradiated for 4 hours, the sizes of primary particles were similar to those of the sample irradiated for 2 hours. However, the image contrast was higher than that of the sample irradiated for 2 hours, implying improved crystallinity. Heavier agglomeration was also found, which may have caused the weakening of quantum size effects and, in turn, the red-shift of the absorption peak.

Although further investigation is necessary to elucidate the mechanism of the UV-light induced room temperature reaction, the possible model is proposed as follows: Firstly, UV-light degrades Zn-TEG chelates to release zinc ions into TEG. It is reported that UV light irradiation can degrade metal chelates to release metal ions (Sun & Pignatello, 1993). Then, the released Zn ions react with OH- ions to form ZnO quantum dots. The newly formed ZnO quantum dots act as a photocatalyst under UV light, by photo-excited electrons and holes creating free radicals when ZnO quantum dots contact with oxygen and water molecules in TEG. The free radicals will decompose the remaining Zn-TEG chelates, releasing more Zn ions to react with OH- ions that increases the number of ZnO quantum dots to be formed. Furthermore, the photo-generated free radicals decompose the TEG on ZnO quantum dots. This weakens the steric hindrance effect of the surface TEG and hence induces particle agglomeration and an increase in crystal size, which results in the red-shift of the optical absorption edge as shown in Figure 8.

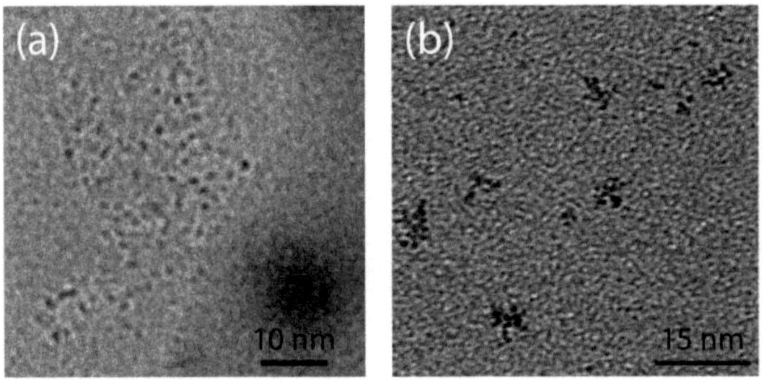

Fig. 9. Bright-field TEM images of ZnO quantum dots synthesized under the UV-irradiation for (a) 2 hours, and (b) 4 hours.

4. Conclusion

In this chapter, novel near-room-temperature methods to synthesize ZnO quantum dots using a polyol were presented. TEG was used as the solvent system in order to enable safe, economical and facile synthesis of free-standing quantum dots. No surfactant or capping agent to limit the particle size growth was added into the reactant mixture. Although $ZnCl_2$ and NaOH can react readily to form $Zn(OH)_2$ precipitates in an aqueous environment, the mixture of $ZnCl_2$ and NaOH solutions in TEG did not induce the reaction at room temperature without external energy input. It was found that mechanical agitation and UV light irradiation can be used to initiate the reaction to form ZnO quantum dots at room temperature. The nanoparticles consisted of ~ 3 - 5 nm sized crystalline primary particles and showed a quantum size effect as a blue-shift of the bandgap energy. It is suggested that

TEG did not only act as a solvent for the reactants but also acted as a surfactant to limit the size of the primary nanoparticles. TEG molecules may bond with ZnO through hydrogen bonding between the -OH group of the surface of ZnO nanoparticles and the -OH terminal of TEG molecules, providing a steric hindrance effect to give high dispersion stability in TEG.

It was proposed that Zn ions form Zn-TEG chelates in TEG and that UV-light irradiation or mechanical milling releases Zn ions from Zn-TEG chelates to form ZnO quantum dots. Further investigation is required to elucidate the reaction mechanism. Extended X-ray absorption fine structure analysis and NMR measurements will allow the conformation of Zn-TEG chelate formation and subsequent release of Zn by UV-irradiation and mechanical milling. The in-depth understanding of the reaction mechanism helps optimise the synthesis conditions to obtain ZnO quantum dots with tailored properties. It is also important to identify the advantages of ZnO quantum dots produced using this method, over the ZnO produced using other methods, in terms of properties and performances in key applications areas such as photocatalysis, gas-sensing, bio-marking and photoluminescence devices.

It is expected that this new synthesis method is applicable for the production of other quantum dots in a safe and scalable manner. In order to further develop the knowledge and techniques on this new method, it is also necessary to investigate the applicability of this method to the synthesis of non-oxide nanoparticles.

5. References

Bauer, C.A.; Timofeeva, T.V.; Settersten, T.B.; Patterson, B.D.; Liu, V.H,; Simmons, B.A. & Allendorf, M.D. (2007). Influence of Connectivity and Porosity on Ligand-Based Luminescence in Zinc Metal−Organic Frameworks. *Journal of the American Chemical Society,* Vol. 129, No. 22, pp. 7136-7144, ISSN 0002-7863

Becheri, A.; Durr, M.; Lo Nostro, P. & Baglioni, P. (2008). Synthesis and characterization of zinc oxide nanoparticles: application to textiles as UV-absorbers. *Journal of Nanoparticle Research,* Vol. 10, No. 4, pp. 679-689, ISSN 1388-0764

Beydoun, D.; Amal, R.; Low, G. & McEvoy, S. (1999). Role of Nanoparticles in Photocatalysis. *Journal of Nanoparticle Research,* Vol. 1, No. 4, pp. 439-458, ISSN 1388-0764

Dobryszycki, J. & Biallozor, S. (2001). On some organic inhibitors of zinc corrosion in alkaline media. *Corrosion Science,* Vol. 43, No. 7, pp. 1309-1319, ISSN 0010-938X

Goharshadi, E.K.; Abareshi, M.; Mehrkhah, R.; Samiee, S.; Moosavi, M.; Youssefi, A. & Nancarrow, P. (2011). Preparation, structural characterization, semiconductor and photoluminescent properties of zinc oxide nanoparticles in a phosphonium-based ionic liquid. *Materials Science in Semiconductor Processing,* Vol. 14, No. 1, pp. 69-72, ISSN 1369-8001

Gronlund, P.J.; Burt, J.A. & Wacholtz, W.F. (1995). Synthesis and characterization of luminescent mixed ligand zinc(II) complexes containing a novel dithiol ligand. *Inorganica Chimica Acta,* Vol. 234, No. 1-2, pp. 13-18, ISSN:0020-1693

He, R.L. & Tsuzuki, T. (2010). Low-Temperature Solvothermal Synthesis of ZnO Quantum Dots. *Journal of the American Ceramic Society,* Vol. 93, No. 8, pp. 2281-2285, ISSN 0002-7820

Hosono, H. (2007). Recent progress in transparent oxide semiconductors: Materials and device application. *Thin Solid Films*, Vol. 515, No. 15, pp. 6000-6014, ISSN 0040-6090

Kumbhakar, P.; Singh, D.; Tiwary, C.S. & Mitra, A.K. (2008). Chemical synthesis and visible photoluminescence emission from monodispersed zno nanoparticles. *Chalcogenide Letters*, Vol. 5, No. 12, pp. 387-394, ISSN 1584-8663

Larcher, D.; Sudant, G.; Patrice, R. & Tarascon, J.M. (2003). Some Insights on the Use of Polyols-Based Metal Alkoxides Powders as Precursors for Tailored Metal-Oxides Particles. *Chemistry of Materials*, Vol. 15, No. 18, pp. 3543-3551, ISSN 0897-4756

Sun, Y.& Pignatello, J.J. (1993). Activation of hydrogen peroxide by iron (III) chelates for abiotic degradation of herbicides and insecticides in water. *Journal of Agricultural and Food Chemistry*, Vol. 41, pp. 308-312.

Tang, X,; Choo, E.S.G.; Li, L.; Ding, J. & Xue, J. (2009). One-Pot Synthesis of Water-Stable ZnO Nanoparticles via a Polyol Hydrolysis Route and Their Cell Labeling Applications. *Langmuir*, Vol. 25, No. 9, pp. 5271-5275, ISSN 0743-7463

Tauc, J.; Grigorovici, R. & Vancu, A. (1966). Optical Properties and Electronic Structure of Amorphous Germanium. *Physica Status Solidi (b)*, Vol. 15, No. 2, pp. 627-637, ISSN 1521-3951

Tong, H.; Umezawa, N. & Ye, J. (2011). Visible light photoactivity from a bonding assembly of titanium oxide nanocrystals. *Chemical Communications*, Vol. 47, No. 14, pp. 4219-4221, ISSN 1359-7345

Tsuzuki, T. & McCormick, P.G. (2001). ZnO nanoparticles synthesised by mechanochemical processing, *Scripta Materialia*, Vol. 44, No. 8-9, pp. 1731-1734. ISSN 1359-6462

Vafaee, M. & Ghamsari, M.S. (2007). Preparation and characterization of ZnO nanoparticles by a novel sol-gel route. *Materials Letters*, Vol. 61, No. 14-15, pp. 3265-3268, ISSN 0167-577X

van Dijken, A.; Meulenkamp, E.A.; Vanmaekelbergh, D. & Meijerink, A. (2000). The kinetics of the radiative and nonradiative processes in nanocrystalline ZnO particles upon photoexcitation. *Journal of Physical Chemistry B*, Vol. 104, No. 8, pp. 1715-1723, ISSN 1089-5647

Wang, Z.L.; Lin, C.K.; Liu, X.M.; Li, G.Z.; Luo, Y.; Quan, Z.W.; Xiang, H.P. & Lin, J. (2006). Tunable photoluminescent and cathodoluminescent properties of ZnO and ZnO : Zn phosphors. *Journal of Physical Chemistry B*, Vol. 110, No. 19, pp. 9469-9476, ISSN 1520-6106

Xiong, H,M,; Liu, D.P.; Xia, Y.Y. & Chen, J.S. (2005). Polyether-Grafted ZnO Nanoparticles with Tunable and Stable Photoluminescence at Room Temperature. *Chemistry of Materials*, Vol. 17, No. 12, pp. 3062-3064, ISSN 0897-4756

Xiong, H.M.; Ma, R.Z.; Wang, S,F, & Xia, Y.Y. (2011), Photoluminescent ZnO nanoparticles synthesized at the interface between air and triethylene glycol. *Journal of Materials Chemistry*, Vol. 21, No. 9, pp. 3178-3182, ISSN 0959-9428

Xiong, H.M. (2010). Photoluminescent ZnO nanoparticles modified by polymers. *Journal of Materials Chemistry*, Vol. 20, No. 21, pp. 4251-4262, ISSN 0959-9428

Zhang, L.; Yin, L.; Wang, C.; Lun, N.; Qi, Y. & Xiang, D. (2010). Origin of Visible Photoluminescence of ZnO Quantum Dots: Defect-Dependent and Size-Dependent. *The Journal of Physical Chemistry C*, Vol. 114, No. 21, pp. 9651-9658, ISSN 1932-7447